Antisense Technology: Methods of Delivery and RNA Studies

Antisense Technology: Methods of Delivery and RNA Studies

Editor: Macy Rose

R CALLISTO REFERENCE

www.callistoreference.com

Callisto Reference,
118-35 Queens Blvd., Suite 400,
Forest Hills, NY 11375, USA

Visit us on the World Wide Web at:
www.callistoreference.com

ISBN: 978-1-64116-785-7 (Hardback)

Cataloging-in-publication Data

Antisense technology : methods of delivery and RNA studies / edited by Macy Rose.
    p. cm.
Includes bibliographical references and index.
ISBN 978-1-64116-785-7
1. Antisense RNA. 2. RNA. 3. Antisense nucleic acids. I. Rose, Macy.
QP623.5.A58 A58 2023
574.873 28--dc23

Table of Contents

Preface

This book aims to highlight the current researches and provides a platform to further the scope of innovations in this area. This book is a product of the combined efforts of many researchers and scientists, after going through thorough studies and analysis from different parts of the world. The objective of this book is to provide the readers with the latest information of the field.

Antisense RNA (asRNA) refers to a type of single-stranded RNA. It hybridizes with protein coding messenger RNA (mRNA) and complements it to inhibit its translation into a protein. Eukaryotes and prokaryotes, both have naturally occuring asRNAs and in them. They can be divided into two types including short and long non-coding RNAs (ncRNAs). asRNAs primarily control the gene expression and are widely utilized as tools for gene knockdown. Antisense technology offers effective experimental methods for examining gene function and gene regulation by modifying gene expression in mammalian cells. Many RNA-based antisense therapies are being developed currently. However, effective delivery of these therapies to target tissues remains the primary barrier in their clinical application. Antisense technology has a wide range of applications in the treatment of numerous diseases such as viral infections, cancer, and immunological, cardiovascular, ocular and neurological diseases. This book contains some path-breaking studies in antisense technology. It consists of contributions made by international experts. This book is a vital tool for all researching or studying antisense RNA as it gives incredible insights into its design and delivery.

I would like to express my sincere thanks to the authors for their dedicated efforts in the completion of this book. I acknowledge the efforts of the publisher for providing constant support. Lastly, I would like to thank my family for their support in all academic endeavors.

Editor

Chemistry of Nucleic Acids Therapeutics: Introduction and History

Michael J. Gait and Sudhir Agrawal

Abstract

This introduction charts the history of the development of the major chemical modifications that have influenced the development of nucleic acids therapeutics focusing in particular on antisense oligonucleotide analogues carrying modifications in the backbone and sugar. Brief mention is made of siRNA development and other applications that have by and large utilized the same modifications. We also point out the pitfalls of the use of nucleic acids as drugs, such as their unwanted interactions with pattern recognition receptors, which can be mitigated by chemical modification or used as immunotherapeutic agents.

Key words Antisense, siRNA, Nucleic acid therapeutics, Oligonucleotides, Toll-like receptors, Pattern recognition receptors, Gapmer, Splice switching

1 Introduction to Synthetic Antisense Oligonucleotides and siRNA

Oligonucleotides are short single-stranded sections of DNA or RNA that contain $2'$-deoxyribo-nucleosides or ribo-nucleosides, respectively, which are linked by $3'-5'$ phosphodiester linkages (Fig. 1a). Antisense oligonucleotides are those that are complementary to a section of naturally occurring RNA, such as an mRNA or a viral RNA, to form Watson–Crick base pairs and to thus inhibit a biological function of that RNA. Zamecnik and Stephenson pioneered this concept in 1978 by utilizing antisense oligodeoxyribonucleotides (ODNs) to bind and inhibit the replication of Rous sarcoma virus (RSV) RNA [1]. This work followed much earlier (1969) studies of De Clercq et al. on interferon induction by synthetic polynucleotides and their phosphorothioate analogues [2] and together these early studies heralded the new field of nucleic acids therapeutics that began to accelerate in the mid to late 1980s.

Many further chemistry developments since then in the use of synthetic oligonucleotide analogues, as outlined below, as well as advances in molecular biology, such as in the newer fields of short

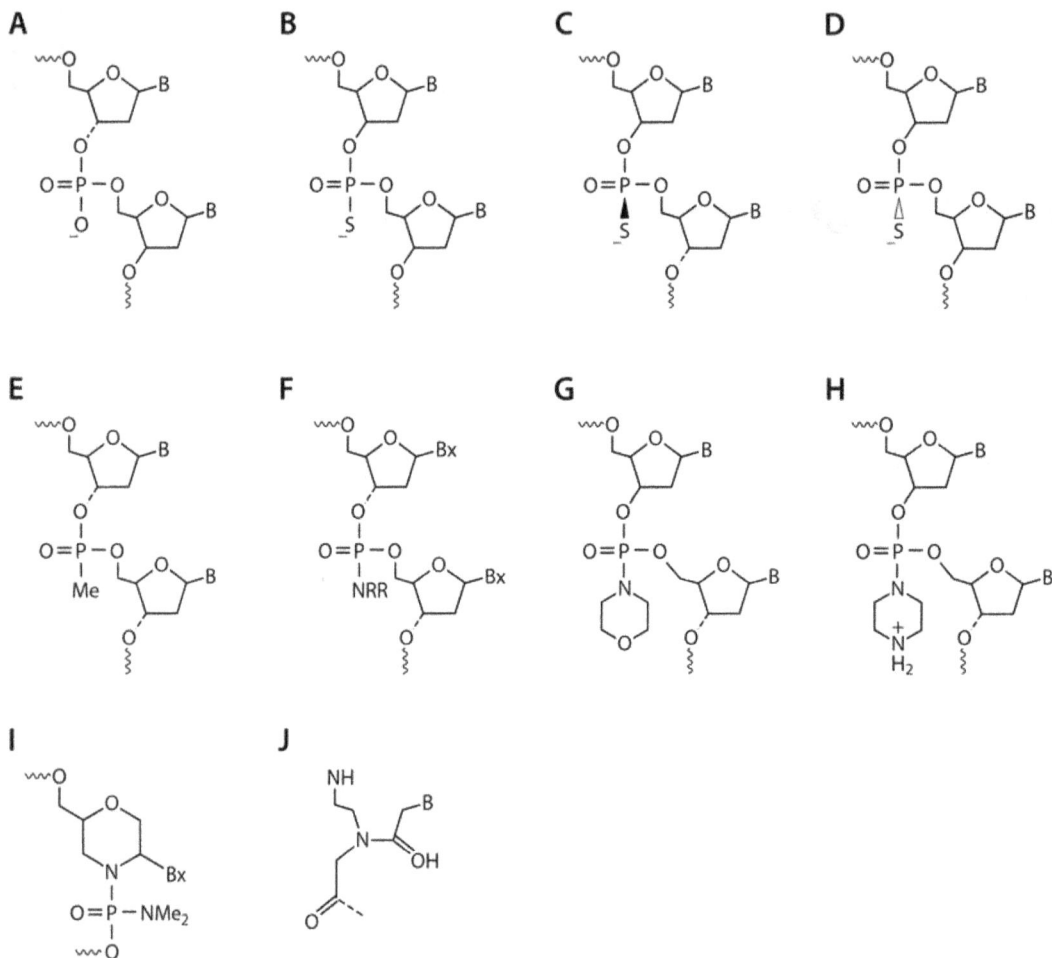

Fig. 1 Chemical structures. (**a**) DNA oligonucleotides (ODN), (**b**) phosphorothioate ODN (PS-ODN), (**c**) *R*p-PS-ODN, (**d**) *S*p-PS-ODN, (**e**) methylphosphonate ODN (PM-ODN), (**f**) phosphoramidate ODN (PN-ODN) R = H or alkyl, (**g**) phosphomorpholidate, (**h**) phosphopiperazidate, (**i**) phosphorodiamidate morpholino (PMO), (**j**) peptide nucleic acids (PNA). B = heterocyclic base

interfering RNA (siRNA) and non-coding RNAs, such as micro-RNA (miRNA), have led to the widespread and convenient use of synthetic oligonucleotides as antisense and siRNA reagents for gene ablation or targeting of non-coding RNA, as well as their use in animals and in humans leading to the approval of 12 drugs to date. In this chapter we outline the history of oligonucleotide chemistry in antisense and siRNA that has led to preclinical studies that have guided their development with drug-like properties and hence clinical trials (Fig. 2). We go on to discuss the development of the principles of widely used antisense gapmers and siRNAs as well as their immune responses by triggering pattern recognition receptors (PRRs) and how such activities can be controlled or harnessed for

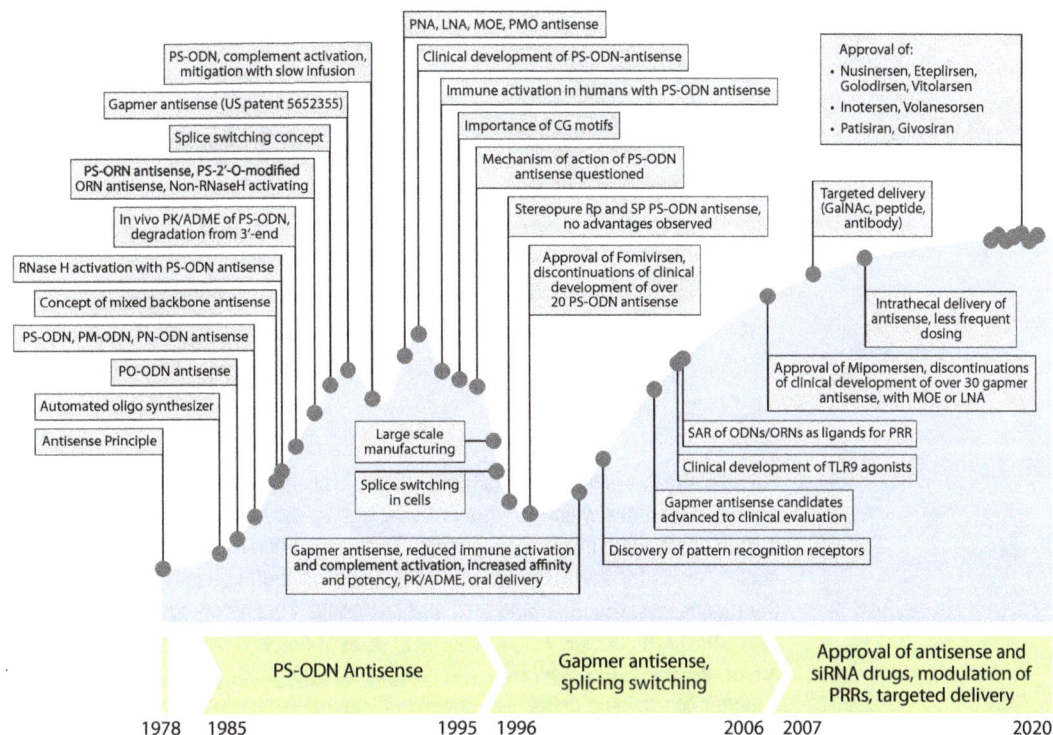

Fig. 2 Evolution of chemical modifications during the development of antisense as therapeutic agents. In the nineties, various modifications of ODNs and ORNs were studied and based on RNase H activation, PS-ODNs became the choice as first-generation antisense agents. Soon it was realized that PS-ODN had off-target activities including complement activation and also sequence specific immune activation. This led to questions on the mechanism of action of PS-ODNs and associated safety signals, and clinical development of most of the PS-ODN ASOs was discontinued. In parallel use of modified ORN for splicing correction in the cells was established. Based on the early work conducted in early nineties, gapmer antisense design provided the key attributes and became the choice as second-generation antisense agents. Studies of chemical modifications in antisense has facilitated development of other therapeutic oligonucleotides. The key modifications which have been identified in the development of antisense, e.g. PS-PDN and PS-ORN, 2'-modified or 2'-O-substituted ribo-nucleosides, bridged ribo-nucleosides, and PMO which are being employed in various nucleic acid therapeutics. In the last few years, a number of drugs based on gapmer ONs (mipomersen, inotersen, volanesorsen), 2'-MOE PS-ORN (nusinersen), PMO (eteplirsen, golodirsen, vitolarsen), and siRNA (patisiran, givosiran) have been approved

use as drugs (Fig. 3). Readers are also referred to a recent book edited by us on recent Advances in Therapeutic Nucleic Acids [3].

2 Oligonucleotide Synthesis

Early work in the 1960s and 1970s on phosphodiester and phosphotriester chemistry for the synthesis of ODNs required armies of nucleic acids chemists for painstaking oligonucleotide synthesis in solution phase that took months to years for each synthesis (for

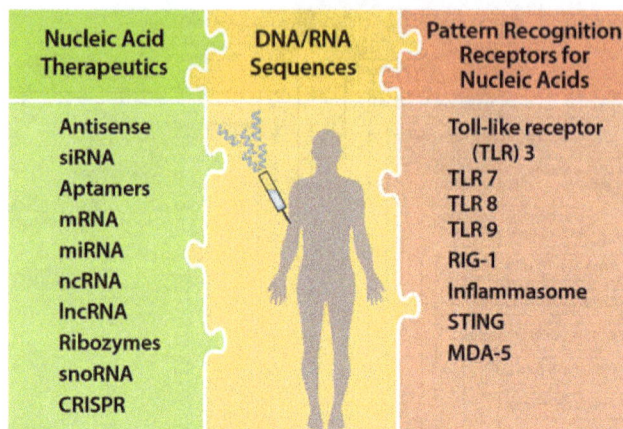

Fig. 3 Pattern recognition receptors (PRRs). The number of PRRs listed in the right-side column which is known to recognize patterns of nucleic acids and induce appropriate immune responses. This recognition is applicable to all use of nucleic acids for therapeutics, and depending on the engagement could affect the mechanism of action and also safety signals. This recognition could be due to PAMPs of the nucleic acid agent being used or due to DAMPs, due to the build-up of administered agent in the tissues and organs, being recognized as endogenous nucleic acids

example [4]). The revolution for molecular biologists came in the late 1970s and early 1980s with the development of solid-phase ODN synthesis first suggested by Letsinger and Mahadevan in 1965 [5] and later developed into working methods in the laboratories of Gait [6–8] and Itakura [9]. These methods were then superseded by the more efficient phosphoramidite chemistry of Caruthers and colleagues [10], which went on to be automated, such as by Applied Biosystems and other companies. The rapid and automated synthesis allowed molecular biologists to obtain synthetic ODNs readily for biological purposes such as for sequencing, cloning, and gene synthesis. The history of oligonucleotide synthesis chemistry has been reviewed [11, 12]. Today standard and modified ODNs can be obtained rapidly and efficiently on a small to large scale through highly automated solid-phase DNA oligonucleotide synthesis for many biological and diagnostic purposes.

Oligoribonucleotide synthesis is also now well established for the synthesis of siRNA or for aptameric RNAs. Currently, several oligonucleotide manufacturing plants are operational to produce oligonucleotides under GMP conditions.

3 Synthetic Oligonucleotide Analogues in Antisense and siRNA

The early work of Zamecnik and colleagues utilized RSV since this was the only viral RNA sequenced at this time. Zamecnik noticed that both ends of the linear RNA genome bore the same primary sequence in the same polarity and that DNA might be synthesized from the RNA by reverse transcription via circularization of the 5′-end with the 3′-end through base pairing. Zamecnik and Stevenson were able to block this circularization by use of a synthetic ODN via hybridization with the 3′-end of the viral RNA. By use of a cell-free system, translation of the RSV mRNA was impaired, thus leading to inhibition of viral replication. This work was the birth of the antisense concept [1, 13].

Further progress in the antisense field awaited the mid to late 1980s for the availability of genomic DNA (or RNA) sequences for antisense targeting as well as the new automated methods of oligonucleotide synthesis as described above. There was also some scepticism regarding the stability and eukaryotic cell entry ability of unmodified oligonucleotides. Nevertheless Zamecnik and Gallo were able to employ unmodified antisense ODNs to inhibit human immunodeficiency virus 1 (HIV-1) replication [13] and to suppress expression of HIV-1 related markers [14]. Cellular uptake of the ODN was not a limiting factor since experiments were carried out in primary human cells and non-targeted control oligonucleotides showed minimal inhibition of HIV-1 replication, thus demonstrating sequence-specificity. This paper reignited the therapeutic potential of the antisense approach.

The next step was to provide drug-like properties to the unmodified antisense ODNs through chemical modifications. In this context the key consideration was to provide nucleolytic stability to antisense ODNs without affecting their hybridization and affinities with the RNA target. Not much was known at the time about the in vivo characteristics of unmodified ODN, or any modified ODNs, which could have guided the study of potential modifications. The first step was to see if modifications of the internucleotide linkages would provide nucleolytic stability to ODNs, while preserving the hybridization affinity to the target RNA. Later on, discovery of PRRs further provided insights into recognition of pathogen associated molecular patterns (PAMPs) of nucleic acids, and how sequence of antisense and nucleic acid-based therapeutic approaches could affect the mechanism of action (Fig. 3).

3.1 Backbone Analogues

3.1.1 Phosphorothioates

The antisense field took inspiration from the very early work of De Clerq, Eckstein, and Merigan [2] where phosphorothioate (PS) modifications were studied in homopolynucleotides as stabilizing agents. By the mid-1980s new chemical synthesis methods for the PS linkage in ODNs became available [15]. Here, a simple sulfur atom replaces a non-bridging oxygen atom (Fig. 1b). However, standard automated synthesis, which in the phosphoramidite method involves treatment with a sulfurizing agent in place of oxidation by iodine, produced a mixture of diastereomeric oligonucleotide products (*R*p and *S*p) (Fig. 1c, d) and thus there was found to be a lower binding affinity to target RNA compared to phosphodiesters. However, PS-linked ODNs are much more resistant to nuclease degradation than phosphodiesters. Optimization of the synthesis methodology allowed the synthesis of milligram quantities of PS-ODNs for use in cell and in vivo experiments.

Early studies showed dose-dependent inhibition of viral replication and antiviral activity in HIV-1 infected cells by use of antisense PS-ODNs targeted to several regions of HIV-1 mRNA [16, 17]. Surprisingly, homopolymers were also effective and antiviral activity depended on the base composition, suggesting that PS-ODNs also had off-target activity. Longer PS-ODNs were more effective than shorter ones and cellular uptake was efficient in primary human cells without a carrier. In addition, antisense PS-ODNs showed potent and durable inhibition of HIV-1 replication in chronically HIV-1 infected cells [18–20]. Soon after, studies with PS-ODN targeted to influenza virus showed inhibition of virus replication [21]. Based on these early studies and promise, PS-ODNs became the choice for first-generation antisense agents.

Following these studies, work on antisense was pursued in many laboratories around the world for a broad range of applications [22–24]. In addition, several new companies were founded to advance therapeutic applications of antisense, such as Gilead Sciences, Isis Pharmaceuticals (now Ionis Pharmaceuticals), Hybridon (now Idera Pharmaceuticals), and others. Numerous reports appeared on the use of antisense PS-ODNs to target viruses [25, 26], oncogenes [24, 27], and kinases [27, 28], etc. Cellular uptake of PS-ODNs in transformed cells in culture was found to be poor but could be improved substantially by use of cationic lipid carriers. It was also shown that an antisense PS-ODN bound to its target RNA engaged RNase H [19, 20] to excise the RNA strand and this was therefore likely to be the mode of action and not steric block inhibition of translation as observed with other modified ODNs [29, 30]. Nevertheless RNase H cleavage activity was poorer than for a PO-ASO [20].

The first in vivo study in mice of a systemically delivered PS-ODN showed that plasma half-life was very short but that there was a broad tissue disposition with most delivered to liver and kidneys and the lowest amounts in the brain [31]. The PS-ODN was stable in tissues for several days and excreted primarily in urine in degraded form mostly through exonuclease cleavage from the 3'-end. Chemical modifications at the 3'-end increased the stability [32, 33]. The PS-ODN was bound by serum proteins, which increased the plasma half-life and improved tissue disposition [34, 35].

Antisense PS-ODNs showed very potent activity in animal models of viral diseases and cancer [36, 37]. However, in some cases a control PS-ODN also showed some activity, leading to the possibility of off-target effects [38]. For example, a PS-ODN targeted to human papillomavirus inhibited papillomavirus-induced growth of implanted human foreskin in a mouse xenograft model but unexpectedly was also active in a mouse cytomegalovirus (CMV) model [39]. Studies in immune-compromised mice showed that the effect of the PS-ODN was largely due to immune activation of the host. Sequence-dependent immune stimulation was confirmed during non-clinical safety evaluations of drug candidates. Repeated systemic administration of PS-ODN candidates in mice and rats caused inflammation, splenomegaly, and histological changes in multiple organs [40, 41]. Further in non-human primates, bolus administration of the first antisense PS-ODN (GEM91) led to severe hemodynamic changes due to activation of the alternative complement pathway [42]. Stimulation of the alternative pathway complement activation cascade became the first documented off-target effect, which was due to a plasma concentration effect of the poly-anionic nature of PS-ODNs. It could be mitigated by subcutaneous administration or by slow intravenous infusion. Thereafter, the US Federal Drug Agency (FDA) published guidelines and required the use of non-human primates for non-clinical safety studies of all oligonucleotide drugs [43].

In the 1990s a number of antisense PS-ODN clinical candidates were advanced to human trials [44] through intravenous infusion, intravitreal or subcutaneous delivery [45–47]. Humans showed similar pharmacokinetics and excretion data to those of non-human primates [48, 49]. However, most clinical studies were discontinued due to the lack of activity or a poor therapeutic index [50]. For example, the subcutaneous administration of GEM91 in humans caused flu-like symptoms, swelling of the draining lymph nodes, prolongation of activated partial thromboplastin time (aPTT), and thrombocytopenia [29]. Rather than suppression of HIV-1, HIV-1 RNA levels were increased in blood [51]. However, intravenous delivery had minimal effect on these parameters. There were a few reports of immunostimulatory properties of DNA/ODNs containing CG nucleotides [52, 53]. It only became clear much later that PS-ODNs containing an unmethylated CpG motif activated the immune responses by binding to Toll-like receptor 9 (TLR9), an innate immune receptor present in immune cells that recognizes DNA containing CpG dinucleotide motifs [54]. It became clear that the flu-like symptoms and injection site reactions seen with most of the PS-ODN drug candidates in clinical trials, such as the clinically approved drug fomivirsen, administered intra-vitreally to treat AIDS-related CMV-induced retinitis, contained a CpG motif [55]. Thus, the true mechanism of action of this first-generation drug, now no longer marketed, remains unclear. Altogether, preclinical and clinical studies have provided important insights into the properties of PS-ODNs and its use as drugs [35, 56, 57].

Also of debate for some time has been whether the presence of a mixture of Rp and Sp diastereoisomers in the synthetic PS-ODNs (Fig. 1c, d) bears any influence on their biological properties. For example, a 20-mer would have 2^{19} isomers. The stereospecificity of enzymes that act on nucleoside phosphates was well known from early work of Eckstein (reviewed in [58]). Since PS-ODN interacting enzymes, such as nucleases, also utilize only a single diasterioisomeric isomer [59], it was plausible that there might be a significant biological effect in cells of utilizing mixed PS diastereomers in antisense PS-ODNs. Testing of this only awaited the solid-phase synthesis of stereo-enriched and stereo-pure PS-ODNs. This became possible through pioneering work of Stec and later by use of nucleoside bicyclic oxazaphospholidinium synthons [15, 60]. It is now known that binding strength and recognition by RNase H is generally higher for antisense oligonucleotides containing Rp linkages but depends crucially on the placement of these with respect to Sp linkages and overall stereospecific PS-ODNs have had limited therapeutic utility [61]. Recently certain stereo-pure antisense oligonucleotides were shown to have improved activity in cell culture and in vivo [62] but the therapeutic significance of such stereospecificity is currently hotly disputed [63]. Even more recently, the clinical development of a stereo-pure PS-ODN, WVEN-531 targeted to DMD has been discontinued due to lack of clinical activity [64]. Furthermore, dosing of this antisense ODN also led to transient increases in complement factors and C-reactive protein [65].

3.1.2 Charge-Neutral Analogues

Two phosphate-containing, charge-neutral oligonucleotide analogues that were particularly used in early antisense studies are the methylphosphonate [66] (Fig. 1e) and the phosphoramidate linkages [67] (Fig. 1f). They both consist of a mixture of diastereoisomers. Methylphosphonate ODNs (MP-ODNs) are stable at physiological pH and are resistant to nucleolytic degradation but are less strongly bound to target RNA compared to PO-ODNs [66, 68]. MP-ODNs targeted to HIV-1 showed dose-dependent inhibition of HIV-1 replication [16], but they are less active than PS-ODNs due to their lack of RNase H activation [19] but instead inhibit protein translation, which is generally a weaker activity in cells. Limited in vivo studies with a MP-ODN showed that while this modification is very resistant to nucleolytic degradation, due to poor protein binding, there was a very poor in vivo disposition and the majority of the administered ODN was eliminated in urine rapidly (Agrawal, unpublished data). In addition, longer MP-ODNs, which bind more strongly to RNA and which are therefore more potent, are poorly soluble under physiological conditions and thus have not been advanced toward clinical trials. By contrast, a 13-mer antisense oligonucleotide containing all phosphoramidate linkages is more soluble. An anticancer agent (GRN163) inhibits the enzyme telomerase [67] and did get into a clinical trial, however, clinical development was discontinued due to lack of clinical activity. In early work, phosphoramidate-linked antisense oligonucleotides (Fig. 1g, h) targeted to HIV-1 showed similar results in cell-based assays to an MP-ODN and were not pursued [16].

Phosphorodiamidate morpholino oligonucleotides (PMOs) are also charge-neutral but here a morpholino ring replaces the sugar unit (Fig. 1i) [69]. PMOs inhibit translation by a steric block mechanism [70] as they are not recognized by RNase H. They are completely resistant to nucleases but are not taken up well by cells and thus require very high doses for in vivo delivery. They were found to be strong antiviral agents, for example, against Ebola, Marburg, and Chikungunya viruses [71]. Three exon skipping PMO drugs, eteplirsen, golodirsen, and vitolarsen designed to induce alternative splicing and restore the reading frame of mutant dystrophin in patients with Duchenne muscular dystrophy (DMD) [72] have been approved but requires the use of high doses (50 mg/kg or higher). Its therapeutic effectiveness, based on biomarkers, is limited [73], but it is a safe drug at the therapeutic dose. PMO and other chemistries used in exon skipping and other steric block activities have been reviewed [74].

Another initially highly promising, charge-neutral analogue are peptide nucleic acids (PNA), where the sugar-phosphate backbone is replaced by aminoethylglycine units linked by amide bonds (Fig. 1j) [75]. PNA binds strongly to target RNA and, like PMO, they are also completely resistant to degradation by nucleases as well as proteases. Also similar to PMO, duplexes with RNA are not recognized by RNase H and thus PNA acts by a steric block mechanism. Antisense PNAs have been broadly studied as anticancer [76, 77], antiviral [78, 79], and antibacterial agents [80, 81] as well as inhibitors of micro-RNAs [82]. However, once again very high doses are needed in in vivo applications, due to poor cellular uptake and unfavorable pharmacokinetics. Poor in vivo biodistribution is a likely reason for why antisense PNAs have not to date found utility as clinical candidates.

3.2 Sugar Analogues

It has been long established that an RNA-RNA duplex is much stronger than that of DNA-RNA. However, RNA (Fig. 4a) is highly unstable to ribonucleases. Phosphorothioate analogues (PS-ORN) showed an increased affinity to target RNA but they were found to have lower potency as compared to PS-ODNs, probably since RNA-RNA duplexes lack RNase H activation ability [35].

3.2.1 2′-O-Alkyl Sugars

The first sugar analogues to find utility in antisense oligonucleotides are the naturally occurring 2′-O-methylribonucleosides (Fig. 4b) first synthesized by the Ohtsuka laboratory [83]. The phosphoramidites of 2′-O-methylribonucleosdes suitable for

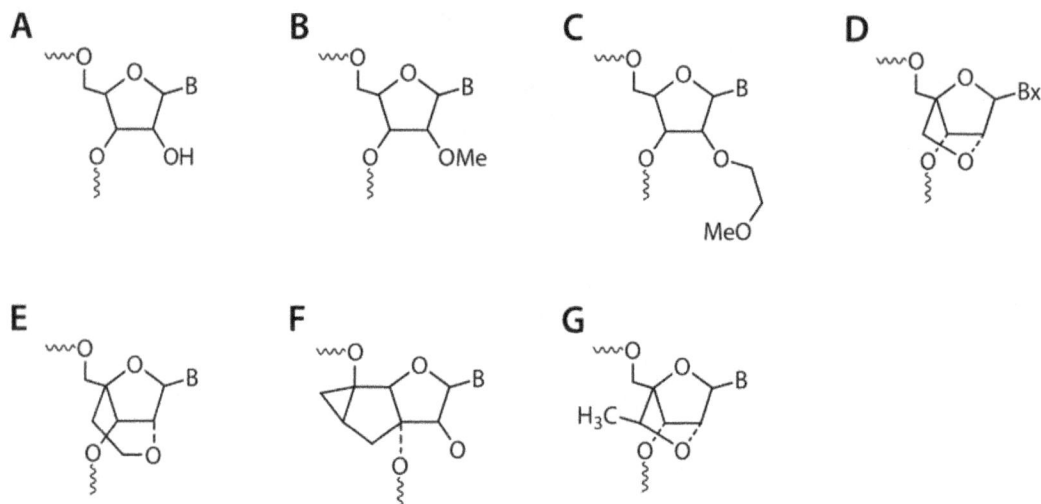

Fig. 4 Chemical structures of the ribo-nucleoside units of therapeutically useful RNA and RNA analogues. (a) ribo-nucleoside (ORN) (b) 2'-O-methyl (2'-OMe), (c) 2'-O-methoxyethyl (2'-MOE), (d) bridged/locked nucleic acid (LNA), (e) 2'-O,4'-C-ethylene linked nucleic acid (ENA), (f) tricyclo-DNA (tcDNA), and (g) constrained ethyl (cET). B = heterocyclic bases

solid-phase synthesis became available commercially in the early 1990s [29]. Studies with 2'-O-methyloligoribonucleotide phosphorothioates (2'-OMe PS-ORN) showed enhanced stability to nucleases as compared to PS-ORN and showed a higher affinity to target [84]. However, they also showed lower antisense activity compared to PS-ODNs, demonstrating that activation of RNase H was key for this activity [30, 51, 85, 86]. Since then, very many additional 2'-O-alkyl analogues have been synthesized and tested in antisense oligonucleotides (ONs), predominantly in gapmer studies (see below) to allow recognition by RNase H. From these studies 2'-O-methoxyethylribonucleoside (2'-MOE) (Fig. 4c) has been employed widely in clinical gapmer candidates (Chapters 3 and 4 of Agrawal and Gait [3]).

As described earlier, dose-dependent activation of complement and prolongation of aPTT were found to be unwanted side effects of PS-ODNs. These effects as well as strong binding to serum proteins were thought to be due to the poly-anionic nature of the PS linkage. However, there were found to be significantly less side effects when PS-ORN or 2'-OMe-PS-ASO where used, suggesting that the poly-anionic nature of the PS backbone in PS-ORN and PS 2'-OMe is not as pronounced when placed in the context of an RNA or RNA-like sugar conformation [40, 85–87]. This became crucial to their use in later gapmer antisense studies.

Uniformly 2'-O-alkyl modified PS-RNA has also found very high therapeutic use in splice switching (exon skipping or exon inclusion) and other steric blocking applications due to their high binding strength to nuclear pre-mRNA [88, 89]. However, the exon skipping 2′-OMe PS-ORN antisense drisapersen drug candidate failed to show clinical benefit in patients with DMD and also caused significant adverse side effects and was, therefore, not approved for clinical use [90]. By contrast, the 18-mer 2′-MOE phosphorothioate (2′-MOE PS-ORN, nusinersen), which redirects the splicing of the SMN-2 gene to generate active SMN protein (exon inclusion), administered intrathecally only few times a year was approved by the FDA for the treatment of spinal muscular atrophy (SMA) [91, 92]. The thrombocytopenia and renal toxicity observed with the use of drisapersen could be largely due to the need for repeated subcutaneous dosages of 2′-OMe PS-ORN, which is very stable to nucleolytic degradation and therefore may accumulate in tissues due to its long half-life and potentially interact with PRRs. Nusinersen, a 2′-MOE PS-ORN, is also quite stable to nucleolytic degradation, but its intrathecal administration and need for infrequent and lower doses minimizes the impact of tissue accumulation and avoids the need for passage into the brain and spinal cord from the circulation.

3.2.2   Locked/Bridged Nucleic Acids

A major step forward in the design of antisense ONs was the development in the laboratories of Wengel and also of Imanishi of bicyclic sugar analogues known as locked or bridged nucleic acids (LNA/BNA). Here the conformational flexibility of nucleotides is significantly reduced by linkage of the 2′-oxygen atom to the 4′-carbon atom in the ribose ring (Fig. 4d). This results in a significant increase in the binding affinity of ONs to complementary RNA targets with an increase in the melting temperature of 2–8 °C per residue [93]. Unfortunately, LNA oligomers of 8 units or longer tend to self-aggregate. Therefore they became more useful as mixmers with 2′-deoxynucleotides and here miravirsen, the first microRNA-targeting drug, which acts by sterically blocking microRNA-122, highly expressed in liver, was developed for the treatment of hepatitis C virus infection, a debilitating liver disease [94]. Unfortunately, this drug's clinical development was discontinued because of *safety* issues. LNA has also been used as mixmers with 2′-OMe nucleotides targeting various RNAs in cells (e.g. [93]) and has also found utility in the flanking sequences of gapmers. This had the effect also of modulating the binding strength of the ON and increasing the specificity of the interaction (reviewed in Chapter 3 of Agrawal and Gait [3]).

Another bicyclic analogue that became useful is 2′-O,4′-C-ethylene linked nucleic acid (ENA) [95] (Fig. 4e). In a recent study an antisense ON DS-5141, containing segments of 2′-OMe PS-RNA and ENA, showed good activity in an *mdx* mouse model of DMD and a phase 1/2 clinical trial was carried out in Japan [96]. A further analogue useful in steric block applications is tricyclo-DNA (tcDNA) [97] (Fig. 4f). However, perhaps the

most important bicyclic derivative of LNA that has found considerable therapeutic utility is the methylated analogue known as "constrained Ethyl" (cET), which is being employed in shorter gapmers
[98] (Fig. 4g) and being evaluated in preclinical and clinical studies
by Ionis Pharmaceuticals Inc. All these types of bridged nucleic
acids have shown very strong affinity to target RNA and increased
nucleolytic stability, but none of them are substrates for RNase H.
Thus, these types of bicyclic sugar analogues are mostly used in
steric block/splicing modulation approaches and in the flanking
sequences of gapmers.

3.3 Heterocyclic Base Analogues

In early antisense studies it was thought that increased antisense
activity might be achievable by improving the affinity of an ODN to
target RNA through modification of the heterocyclic bases, for
example, by adding an extra hydrogen bond in the base pairing
between an ODN and its RNA target or by increasing the base
stacking potential in a DNA–RNA duplex. Chemically this was
simplest through modifications in the pyrimidine rings, for example, by modifications at positions C-2, C-4, C-5 or at C-6, and
many of these base analogues were incorporated into antisense
ODNs. However, few of these proved to be of significant value.
Incorporation of modified purines generally resulted in a reduced
binding affinity of an antisense ODN. Perhaps the most useful
study of antisense activity was of incorporation of various heterocyclic bases in ODNs including the increased base stacking analogues
C-5 propynyl and 5-methyl cytosine (5-MeC) and the increased
hydrogen-bonding analogues phenoxazine, and G-clamp. These
studies showed that the increased hydrogen-bonding analogue
G-clamp had potent dose-dependent antisense activity [99]. Unfortunately, these antisense ODNs containing G-clamps were found to
be highly toxic in in vivo studies. Currently, the only significantly
used nucleoside base analogue in antisense ODNs is 5-methyl-
2'-deoxycytidine (5-MedC) [100]. This methylated base analogue
is used mainly to mitigate immune activation in CpG dinucleotide
sequences rather than for changing binding strength [101].

4   RNase H Active Gapmer Chemistry for Use as Drugs

Early studies conducted with various modified ODNs, ORNs, and
2'-substituted ORNs as antisense agents provided great insights
into what is important for providing drug-like properties to antisense oligonucleotides [3]. Studies with PS-ODNs showed that
increased nucleolytic stability and activation of RNase H were key.
However, polyanion-related side effects and sequence-dependent
immune activation were limiting factors in their broad applicability
[50, 57]. Studies with MP-ODNs showed that polyanion-related
side effects could be completely mitigated (Agrawal, unpublished

data), and had significant nucleolytic stability. However, with lower affinity and lack of RNase H activation, there was a loss in antisense potency [16]. These observations led to the concept of combining desirable properties of the two modified ODNs to provide drug-like properties to antisense oligonucleotides [29]. The first studies were carried out with antisense containing segments of PS-ODN and PM-ODN or PN-ODN referred to as mixed backbone antisense ONs. These antisense designs showed increased nucleolytic stability and RNase H activation [19, 40]. However, reduced affinity limited their potency. Further insight was obtained from in vivo studies in which a mixed backbone ON containing PS-ODN and PM-ODN showed wide tissue disposition and increased stability and longer half-life in tissues [102].

This led Agrawal and colleagues to design antisense oligonucleotides in which the segments of PS-ODN and 2'-substituted PS-ORN were combined at the appropriate positions [29, 84, 85]. These types of antisense oligonucleotides were referred to as Hybrid ONs, now commonly referred to as gapmers (Fig. 5). In the original design of antisense, a segment of PS-DNA was placed in the middle and segments of 2'-O-alkyl PS-ORN or a combination of PO- and PS-linkages were placed at both 3'- and 5'-ends [29, 84, 85, 87, 103]. This design of antisense combined the desirable properties of PS-DNA and 2'-O-alkyl PS-ORN, and provided increased affinity to targeted RNA, activation of RNase H, increased nucleolytic stability, and reduced polyanion-related side effects. Furthermore, inflammatory responses were also reduced [35, 40, 56]. In vivo administration in mice showed similar plasma half-life and tissue disposition similar to that observed with PS-ODNs, and with increased in vivo stability and retention in tissues [104]. Also, due to increased in vivo stability, oral and rectal delivery of gapmer antisense was possible [34]. It was postulated that the increased stability and in vivo persistence may allow less frequent dosing to obtain therapeutic benefits.

There was also a concern that increased retention of gapmer antisense in tissues may lead to tissue build up following repeated dosing, which would induce local inflammatory responses and side effects, thereby limiting its therapeutic potential.

Other configurations of gapmer antisense were also evaluated, including the configuration in which a segment of 2'-O-alkyl PS-ORN was placed in the center and segments of PS-DNA were placed at both 3'- and 5'-ends. This design of gapmer antisense showed increased potency compared to PS-DNA and reduced polyanion-related side effects. In general, the specificity of RNase H mediated cleavage and its efficiency and excision sites were dependent on the position of the PS-DNA in gapmer antisense ONs [105].

Based on these encouraging results, gapmer antisense became the choice for second-generation antisense agents. In 2001, a

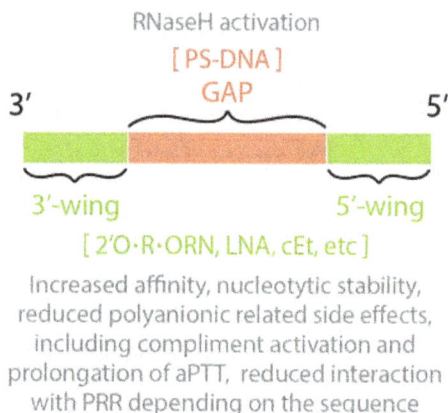

Fig. 5 Design of gapmer antisense oligonucleotides. In a gapmer antisense, segments of PS-DNA and modified RNA are appropriately placed to combine desirable characteristics for antisense agent with both of these modifications. PS-DNA segment provides RNase H activation, and modified RNA segments provide increased nucleolytic stability, affinity to target RNA, decreased polyanionic characteristics and inflammatory responses

licensing agreement between the companies allowed the technology to be widely available [44]. Over the years, studies have been carried out to establish the optimal size of the window of a central PS-DNA segment [105]. Similarly, studies have been carried out to optimize the size of the modified ORN wings at both 3′- and 5′-ends. In the wings of the gapmer antisense, various modified ORNs have been incorporated and evaluated (*see* Chapter 3 of Agrawal and Gait [3]). To date, most promising results have been obtained with gapmer antisense containing segments of 2′-*O*-methyl or 2′-*O*-methoxyethyl at both 3′- and 5′-ends. Over 30 gapmer antisense drug candidates containing 2′-*O*-methoxyethyl or LNA segments have been advanced to clinical evaluations following systemic delivery. To date, three candidates have been approved for clinical use. These include inotersen [106], volanesorsen [107], and mipomersen [108]. Clinical development of several gapmer antisense drug candidates have been discontinued, due to lack of clinical activity and or safety signals. These include ISIS-FXIRx, ISIS-EIF4ERx, ATL1103, ATL1102, ISIS-GCGRRx, ISIS-PTBRx, ISIS-APOARx, ISIS-SOD1Rx, ISIS-FGFR4Rx, ISIS-405879, OGX-011, OGX-427, LY2181308, ATL1103, ATL1102, etc.

As discussed above, many of the bicyclic sugar analogues including locked/bridged nucleic acids have been studied as antisense agents. These analogues have also been studied as part of the wings in gapmer antisense. These include LNA (Fig. 2d) [109], constrained ethyl 2′-4′ bridged nucleic acid (cEt) (Fig. 2g) [110], anhydrohexitol [91], fluorocyclohexenyl (F-CeNA) [111], altritol nucleic acids [112], and tricyclo-DNA (tcDNA) (Fig. 2f) [97]. These modifications provide very high affinity and have allowed the length of the gapmer antisense to be reduced. A few of these shorter gapmer ASOs are being employed to achieve allele specific knockdown [113]. While the use of shorter antisense may be cost effective, it increases the possibility of off-target effects by binding to non-targeted RNA [114, 115]. Also selected LNA and cET ASOs have been associated with liver toxicity [116, 117].

Several gapmer antisense drug candidates employing LNA in the wings have advanced toward the clinic but development of most of these candidates have been discontinued, primarily due to safety issues and lack of therapeutic index.

Other modifications in the gapmer antisense studies include 2′-deoxy-2′-fluoro-beta-D-arabinonucleic acid (FANA) [118], 3′-fluorohexitolnucleic acid (FHNA) [119], 2′-thiothymine, 5-modified pyrimidine bases, etc. These studies are limited to pre-clinical evaluations.

5 siRNA Chemistry for Use as Drugs

The lessons learned in the development of antisense ONs have allowed the development of siRNA therapeutics to be speeded up. siRNAs have a well-defined structure: a short double stranded RNA of 20–25 base pairs with phosphorylated 5′-ends and hydroxylated 3′-ends and also usually containing two 3′-overhanging nucleotides, although blunt ends are sometimes used. While the key requirement is to provide nucleolytic stability to a siRNA candidate, it requires an understanding of the function of each strand. One strand is called the passenger strand and the other is the active component and is called the antisense or guide strand. It is the guide strand that is incorporated into the enzyme complex called RISC in order to be directed to cleave the target RNA strand, while the passenger strand is displaced.

Studies of various chemical modifications in antisense and their impact on providing drug-like properties have allowed the use of some of these modifications in development siRNA therapeutics. These include PS-linkages, various modified ribose sugars such as 2′-O-methyl, 2′-fluoro 2′-deoxy (2′-F), LNA as well as the sugar ring-opened analogues unlocked nucleic acid (UNA), and glycol nucleic acid (GNA). In siRNA chemical modifications are introduced strategically to provide nucleolytic stability. In addition the passenger strand is usually heavily modified in order to block passenger strand entry into RISC, while to promote RISC loading of the guide strand only light modification is used, such as 2′-F replacement of 2′-OH groups in pyrimidines. At the same time modifications must not be placed centrally in the guide strand so as to block RISC-associated cleavage of the target RNA. The exact locations of such modifications in guide and passenger strands are generally closely guarded secrets by reagent suppliers. In addition, the 5'-phosphate of a siRNA guide strand is essential for recognition by RISC. Phosphatase-resistant analogues of the 5'-end phosphate have been shown to improve the in vivo efficacy and are used in clinical candidates [120].

In siRNA candidates, chemical modifications provide nucleolytic stability, however, delivery to a desired tissue or cell type requires use of carrier or conjugation with delivery moieties [121]. To date, two main delivery platforms—ionisable lipid nanoparticles (iLNPs) and trivalent N-acetylgalactosamine (GalNAc) conjugates—have been employed for delivery to liver hepatocytes. To date, two siRNA drugs have been approved for clinical use patisiran, which uses lipid delivery, and givosiran, which uses a GalNAc conjugation. Third siRNA drug candidate, inclisiran, a GalNac conjugate has shown positive results in phase 3 trial [122]. Details of these structure activity relationship studies have been discussed in two chapters from a previous book [123, 124].

6   Immune Responses to Nucleic Acids

Over the last five decades there have been several reports on observations that certain nucleic acid sequences showed immune stimulatory properties [52, 125, 126]. In the mid-1990s subcutaneous administration of an antisense PS-ODN targeted to HIV-1 (GEM91) in HIV-1 infected individuals caused flu-like symptoms and systemic immune responses [51]. This observation alone could not be explained until the discovery of PRRs. These receptors are part of the immune system and PAMPs and host-derived damage-associated molecular patterns (DAMPs). These PRRs play an essential role in establishing antiviral and antibacterial responses by recognizing PAMPs. However, PRRs could also induce development of autoimmune and inflammatory diseases by recognizing DAMPs [127, 128].

PAMPs are highly conserved motifs in pathogens, such as bacteria and viruses. There are several PRRs that are known to recognize motifs, sequences, and patterns of nucleic acids and induce receptor-mediated immune responses. These include members of Toll-like receptors (TLRs), of which four TLRs respond to nucleic acids. TLR3, TLR7/8, and TLR9 recognize double-stranded RNA (dsRNA), single-stranded RNA with certain sequence composition and modified bases (ssRNA), and DNA containing unmethylated CpG sequences (CpG DNA), respectively (Chapter 13 of Agrawal and Gait [3]; [129]). These TLRs are localized in endosomes and expressed on various cell types. The type of immune response induced varies dependent on the receptor and the nature of the nucleic acid [130]. In addition to TLRs, additional receptors are present in the cytoplasm known to recognize nucleic acid-based PAMPs. These include retinoic acid-inducible gene-I (RIG-I), melanoma-associated gene-5 (MDA-5), absent in melanoma 2 (AIM2), cyclic-AMP synthase (cGAS), and stimulator of the interferon gene (STING) (Chapter 13 of Agrawal and Gait [3]).

Following the discovery of TLR9, it became clear that immune activation observed with administration of GEM91 was due to the presence of unmethylated CpG dinucleotides in the antisense sequence [131]. This also provided insights into many of the preclinical studies that the chosen antisense may be exerting antiviral or anticancer activity due to immune activation and not by an antisense mechanism [51, 57]. Interestingly, most of the antisense PS-ODNs in clinical development contained unmethylated CpG motifs, raising questions on the intended mechanism of action [51, 57]. Clinical development of all these antisense drug candidates was discontinued due to lack of activity but also due to safety signals. Similar observations have been made with a few initial siRNA candidates and once again mechanisms of action have been correlated with activation of immune responses [132, 133].

The discovery of PRRs has provided key insights into many of the observations made with use of PS-ODN antisense. For example, TLR9 is a receptor for synthetic ODNs containing unmethylated CpG motifs [131]. Activation of TLR9 leads to induction of Th1 type immune responses in mice, primates, and in humans (Chapter 14 of Agrawal and Gait [3]). Inductions of Th1 type immune responses, which include type interferon (IFN) and interleukin 12 (IL-12), have shown therapeutic potential as antiviral and anticancer agents. This explains the activity of a PS-ODN antisense containing the CpG motif targeted to HPV, also showing activity for CMV and loss of activity in immune compromised mice [39]. This also explains the reason for flu-like symptoms with administration of GEM91, a PS-ODN antisense containing the CpG motif target to HIV-1 [55]. Interestingly, most of the PS-ODN antisense that were advanced to clinical development in the early 1990s contained unmethylated CpG motifs [51]. Thus, their mechanisms of action could be largely due to immune activation or side effects were caused by immune activation.

Detailed structure activity relationships have been carried to elucidate the interaction of PS-ODNs with TLR9. These studies have provided great insights. For example, (a) the presence of unmethylated CpG motif is required, although its position in the sequence is equally important [134], (b) accessibility of the 5'-end is required [111, 135], (c) modifications of the flanking sequence on the 5'-end impacts the immune activity [136], (d) methylation of C in the CpG motif neutralizes immune activation and causes it to act as an antagonist, and (e) certain modified bases could be used in the CpG motif without inducing immune responses. These insights have been very helpful in designing antisense candidates. These lessons have provided the basis for the creation of optimized agonists and antagonists of TLR9. These classes of compounds have been studied extensively in preclinical models of cancer [106], vaccines [137], viral infection [138], and autoimmune diseases [139], and clinical proof of concept has been established in multiple diseases [3]; Chapters 5 and 14 of Agrawal and Gait [3].

Detailed structure–activity relationship studies have been carried out for TLR3 [134], TLR7 and TLR8 [127, 140], RIG-I [141], and AIM2 [142]. It is important to take these insights into consideration when selecting a sequence and prioritizing chemical modification for use in therapeutic applications.

7 Conjugates and Delivery

The in vivo efficacy of ONs is defined by plasma half-life, tissue uptake, nucleolytic stability, and elimination. Systemic administration of several gapmer ASOs has shown a similar profile, i.e., short plasma half-life, wide tissue dispositions, and the presence of intact ASO for longer durations [121]. Even though the delivered gapmer ASO is present in targeted tissues including liver, for sustained clinical activity weekly dosing has been employed. This suggests that the administered ASO is not present in the right cells or cell compartment. Further insights came from intrathecal delivery of the $2'$-O-methoxyethyl PS-ASO nusinersen (Spinraza) to treat SMA. Patients are being treated with IT delivery, administered only four to five times a year. This suggests that in a local compartment, a delivered ASO exerts pharmacodynamic activity for a longer duration and thereby requires less frequent dosing. In recent studies, both preclinical and clinical, conjugation of gapmer ASO with a GalNac cluster has been shown to improve potency and frequency of treatment for liver targeted RNA/gene targets. Efforts are being made to improve delivery of ASOs to muscles to treat muscular disorders employing antibody conjugates [143, 144].

Peptide conjugation has been researched extensively in recent years in efforts to increase the delivery of oligonucleotides. Numerous cell-penetrating peptides (CPPs) have been developed, which are beyond the scope of this Introduction. Readers are referred to a book describing methods that use cell-penetrating peptides [145]. However, the only peptides that have reached clinical development are Arginine-rich CPPs. These are not suitable for conjugation with negatively charged oligonucleotides because of the tendency of such conjugates to aggregate due to charge–charge interactions between the positively charged peptide part and the negatively charged oligonucleotide part. Instead they have found clinical utility for use as conjugates with charge-neutral PMOs.

Here the company AVI Biopharma (now called Sarepta) developed a series of Arginine-rich CPPs that were taken to toxicological testing in monkeys but were found to have renal toxicity at elevated doses leading to a poor therapeutic index [146]. Recently Sarepta has advocated use of an alternative and shorter Arg-rich peptide, which is $(Arg)_6$-Gly, as a PMO conjugate as a treatment for the neuromuscular disease DMD. This gave rise to significant improvements in delivery of an attached PMO and increased exon skipping [147]. The peptide-PMO conjugate is currently in Phase 2 clinical trials. Similar Arg-rich peptides, eg ones known as Pip having a short internal hydrophobic domain, have given rise to increased exon skipping for an attached PMO in muscles as well as in heart in an *mdx* mouse model of DMD [148]. Pip peptides and similar derivatives are currently being evaluated as potential therapeutics for other neuromuscular diseases, for example, in myotonic dystrophy [149]. Once again, shorter Arg peptide derivatives as PMO conjugates are likely to be the future clinical candidates in neuromuscular and neurodegenerative diseases.

Delivery of siRNA has been facilitated by the use of lipid complexes or as conjugates with GalNAc, mainly to the liver. Use of lipids provides stability to an siRNA candidate by encapsulating them, along with preferential delivery to the liver. Several lipid encapsulated siRNA candidates have advanced to clinical development. For example, patisiran formulated with lipid nanoparticles (LNP) has received regulatory approval. It is important to note that lipid-nucleic acid mixtures form complexes that create virus-like particle structures and engage PRRs to induce immune responses. In the case of patisiran, subjects were pre-treated with steroids to mitigate inflammatory responses.

The application of the use of GalNAc to hepatocytes has been known for some time and was employed for oligonucleotide delivery more than two decades ago [150]. GalNAc is a ligand for the asialoglycoprotein receptor 9 (ASGPR), which is very abundant on the surface of hepatocytes [151]. Conjugation with GalNAc generally leads to preferential delivery to the liver. However, depending on the nature of modifications of ONs, delivery to other compartments including the kidney has been observed. An siRNA-GalNAc conjugate givosiran has been approved for clinical use, another GalNAc conjugate. Inclisiran has shown positive results in a phase 3 clinical trial (Chapter 11 of Agrawal and Gait [3]).

8   Further Developments in Therapeutics

Over the years many other applications of nucleic acid-based therapeutics have been pursued. These include aptamers, CRISPR/Cas9, and use of modified mRNA for protein overexpression. While the construct and sequence of DNA or RNA employed in these uses may differ, one common aspect is the need to provide drug like properties to the selected agent. Most of the lessons learned in the development of chemistry in antisense field have facilitated the development of these approaches. In aptamers, modified nucleosides, such as 2'-O-methyl, 2'-fluoro or 2'-amino, and modified internucleotide linkages such as PS-linkages or boranophosphate are regularly employed [152, 153]. In the case of mRNA therapy, considerations of use of chemical modifications are different than in other approaches. The 5'-cap and 3'-poly(A) tail are the key contributors to provide long half-life and for efficient translation. New capping agents such as 1,2-dithiodiphosphonate modified caps have been shown to improve RNA translation [154]. Several modified nucleosides including N^1-methyl-pseudouridine and others have been useful in increasing efficiency of translation, and also mitigating immune stimulatory activity [155, 156]. The positional incorporation of modified bases in mRNA affects the secondary structure of the mRNA, which in turn influences its translation. Further stability to mRNA is provided by formulation with LNPs [157].

In studying CRISPR/Cas9-based therapeutic applications, several modifications are being evaluated. These include PS linkages and 2'-fluoro, LNA, c-Et [158], 2'-O-methyl [159], 2',4'-BNA (NC) [N-Me] [160], etc. These chemical modifications not only provide stability but also mitigate interactions with PRRs. CRISPR-based technologies have been described in a recent book [161].

9   Summary

Nucleic acid-based therapies are now entering into their fifth decade (*see* Fig. 2 for a timeline of developments). Since the first report of the antisense principle in 1978 using unmodified ODNs, the technology has evolved, and drugs are now being approved. Based on the progress to date and the promise of the results, nucleic acid therapeutics are now being recognized as the third major drug discovery and development approach in addition to small molecules and protein/antibody approaches.

Nucleic acid therapeutic agents are built of A, C, G, T, and U nucleotides and connected through internucleotide bonds. Early work on chemical modifications to provide drug-like properties to antisense and lessons learned have been of tremendous value not only in creating antisense drugs but also in developing therapeutics using synthetic nucleic acids with other mechanisms of action (Fig. 4). Nucleic acid therapeutics could be broadly divided into two classes, the first in which an agent is created to target RNA or DNA and modulate its expression, and in the second an agent is created to bind to proteins or cellular factors. In both of these categories, agents could be recognized by PRRs thereby inducing immune responses, either unintended or intended affecting the mechanism of action.

The work on the chemistry of antisense has provided us with a few key modifications that have become important tools in nucleic acid therapeutics. The most important of these include PS linkages in ODN and ORN, gapmer design, selected 2'-O-sustituted nucleosides, and various bridged/locked nucleic acids, etc. The art of creating a nucleic acid agent lies in the understanding of putting together the nucleotide sequence and various modifications for its intended mechanism of action without interacting with PRR (Fig. 3).

Acknowledgments

SA is indebted to Mike Gait for his mentorship during his postdoctoral training in Mike's laboratory and over the last three decades. SA is also grateful to all the colleagues and collaborators whose names appear in the references cited from his laboratory in this chapter.

References

1. Zamecnik PC, Stephenson ML (1978) Inhibition of Rous sarcoma virus replication and cell transformation by a specific oligodeoxynucleotide. Proc Natl Acad Sci U S A 75: 280–284

2. De Clercq E, Eckstein F, Merigan TC (1969) Interferon induction increased through chemical modification of a synthetic polynucleotide. Science 165:1137–1139

3. Agrawal S, Gait MJ (eds) (2019) Advances in nucleic acid therapeutics. Drug discovery series. Royal Society of Chemistry, London

4. Sekiya T, Takeya T, Brown EL, Belagaje R, Contreras R, Fritz H-J, Gait MJ, Lees RG, Ryan MJ, Khorana HG, Norris KE (1979) Total synthesis of a tyrosine suppressor tRNA gene (16). Enzymatic joings to form the total 208 base-pair long DNA. J Biol Chem 254:5787–5801

5. Letsinger RL, Mahadevan V (1965) Oligonucletiode synthesis on a polymer support. J Am Chem Soc 87:3526–3527

6. Gait MJ, Sheppard RC (1976) A polyamide support for oligonucleotide synthesis. J Am Chem Soc 98:8514–8516

7. Gait MJ, Singh M, Sheppard RC, Edge M, Greene AR, Heathcliffe GR, Atkinson TC, Newton CR, Markham AF (1980) Rapid synthesis of oligodeoxyribonucleotides IV. Improved solid phase synthesis of oligodeoxyribonucleotides through phosphotriester intermediates. Nucl Acids Res 8: 1080–1096

8. Sproat BS, Gait MJ (1984) Solid-phase synthesis of oligodeoxyribonucleotides by the phosphotriester method. In: Gait MJ (ed) Oligonucleotide synthesis: a practical approach. IRL Press, Oxford, pp 83–114

9. Miyoshi K, Itakura K (1979) Solid phase synthesis of nonadecathymidylic acids by the phosphotriester method. Tetrahedron Lett 20:3635–3638

10. Beaucage SL, Caruthers MH (1981) Deoxynucleoside phosphoramidites-A new class of key intermediates for deoxypolynucleotide synthesis. Tetrahedron Lett 22:1859–1862

11. Reese CB (2002) The chemical synthesis of oligo- and poly-nucleoties: a personal commentary. Tetrahedron 58:8893–8920

12. Reese CB (2005) Oligo- and poly-nucleotides: 50 years of chemical synthesis. Org Biomol Chem 3:3851–3868

13. Barker RH Jr, Metelev V, Rapaport E, Zamecnik P (1996) Inhibition of Plasmodium falciparum malaria using antisense oligodeoxynucleotides. Proc Natl Acad Sci U S A 93(1):514–518

14. Zamecnik PC, Goodchild J, Taguchi Y, Sarin PS (1986) Inhibition of replication and expression of human T-cell lymphotropic virus type III in cultured cells by exogenous synthetic oligonucleotides complementary to viral RNA. Proc Natl Acad Sci U S A 83(12): 4143–4146

15. Stec WJ, Zon G, Egan W, Stec B (1984) Automated solid-phase synthesis, separation, and stereochemistry of Phosphorothioate analogues of oligodeoxyribonucleotides. J Am Chem Soc 106:6077–6079

16. Agrawal S, Goodchild J, Civeira MP, Thornton AH, Sarin PS, Zamecnik PC (1988) Oligodeoxynucleoside phosphoramidates and phosphorothioates as inhibitors of human immunodeficiency virus. Proc Natl Acad Sci U S A 85(19):7079–7083

17. Matsukura M, Shinozuka K, Zon G, Mitsuya H, Reitz M, Cohen J, Broder S (1987) Phosphorothioate analogs of oligodeoxyribonucleotides: inhibitors of replication and cytopathic effects of human immunodeficiency virus. Proc Natl Acad Sci U S A 84:7706–7710

18. Agrawal S, Ikeuchi T, Sun D, Sarin PS, Konopka A, Maizel J, Zamecnik PC (1989) Inhibition of human immunodeficiency virus in early infected and chronically infected cells by antisense oligodeoxynucleotides and their phosphorothioate analogues. Proc Natl Acad Sci U S A 86:7790–7794

19. Agrawal S, Mayrand SH, Zamecnik PC, Pederson T (1990) Site-specific excision from RNA by RNase H and mixed-phosphate-backbone oligodeoxynucleotides. Proc Natl Acad Sci U S A 87(4):1401–1405

20. Walder RY, Walder JA (1988) Role of RNase H in hybrid-arrested translation by antisense oligonucleotides. Proc Natl Acad Sci U S A 85(14):5011–5015

21. Leiter JM, Agrawal S, Palese P, Zamecnik PC (1990) Inhibition of influenza virus replication by phosphorothioate oligodeoxynucleotides. Proc Natl Acad Sci U S A 87(9): 3430–3434

22. Gaudette MF, Hampikian G, Metelev V, Agrawal S, Crain WR (1993) Effect on embryos of injection of phosphorothioate-modified oligonucleotides into pregnant mice. Antisense Res Dev 3:391–397

23. Knorre DG, Vlassov VV (1991) Reactive oligonucleotide derivatives as gene-targeted biologically active compounds and affinity probes. Genetica 85:53–63

24. Ratajczak MZ, Kant JA, Luger SM, Hijiya N, Zhang J, Zon G, Gewirtz AM (1992) In vivo treatment of human leukemia in a scid mouse model with c-myb antisense oligodeoxynucleotides. Proc Natl Acad Sci U S A 89(24): 11823–11827

25. Cowsert LM, Fox MC, Zon G, Mirabelli CK (1993) In vitro evaluation of phosphorothioate oligonucleotides targeted to the E2 mRNA of papillomavirus: potential treatment for genital warts. Antimicrob Agents Chemother 37(2):171–177

26. Flores-Aguilar M, Freeman WR, Wiley CA, Gangan P, Munguia D, Tatebayashi M, Vuong C, Besen G (1997) Evaluation of retinal toxicity and efficacy of anti-cytomegalovirus and anti-herpes simplex virus antiviral phosphorothioate oligonucleotides ISIS 2922 and ISIS 4015. J Infect Dis 175:1308–1316

27. Monia BP, Johnston JF, Geiger T, Muller M, Fabbro D (1996) Antitumor activity of a phosphorothioate antisense oligodeoxynucleotide targeted against C-raf kinase. Nat Med 2(6):668–675

28. Dean NM, McKay R (1994) Inhibition of protein kinase C-alpha expression in mice after systemic administration of phosphorothioate antisense oligodeoxynucleotides. Proc Natl Acad Sci U S A 91(24): 11762–11766

29. Agrawal S (1992) Antisense oligonucleotides as antiviral agents. Trends Biotechnol 10(5): 152–158

30. Monia BP, Lesnik EA, Gonzalez C, Lima WF, McGee D, Guinosso CJ, Kawasaki AM, Cook PD, Freier SM (1993) Evaluation of 2'-modified oligonucleotides containing 2'-deoxy gaps as antisense inhibitors of gene expression. J Biol Chem 268:14514–14522

31. Agrawal S, Temsamani J, Tang JY (1991) Pharmacokinetics, biodistribution, and stability of oligodeoxynucleotide phosphorothioates in mice. Proc Natl Acad Sci U S A 88(17):7595–7599

32. Temsamani J, Roskey A, Chaix C, Agrawal S (1997) In vivo metabolic profile of a phosphorothioate oligodeoxyribonucleotide. Antis Nucl Acid Drug Dev 7(3):159–165. https://doi.org/10.1089/oli.1.1997.7.159

33. Temsamani J, Tang JY, Padmapriya A, Kubert M, Agrawal S (1993) Pharmacokinetics, biodistribution, and stability of capped oligodeoxynucleotide phosphorothioates in mice. Antisense Res Dev 3(3):277–284

34. Agrawal S, Temsamani J, Galbraith W, Tang J (1995) Pharmacokinetics of antisense oligonucleotides. Clin Pharmacokinet 28(1):7–16.

https://doi.org/10.2165/00003088-199528010-00002

35. Agrawal S, Zhang X, Cai Q, Kandimalla ER, Manning A, Jiang Z, Marcel T, Zhang R (1998) Effect of aspirin on protein binding and tissue disposition of oligonucleotide phosphorothioate in rats. J Drug Target 5(4):303–312. https://doi.org/10.3109/10611869808995883

36. Dean N, McKay R, Miraglia L, Howard R, Cooper S, Giddings J, Nicklin P, Meister L, Ziel R, Geiger T, Muller M, Fabbro D (1996) Inhibition of growth of human tumor cell lines in nude mice by an antisense of oligonucleotide inhibitor of protein kinase C-alpha expression. Cancer Res 56(15):3499–3507

37. Moriya K, Matsukura M, Kurokawa K, Koike K (1996) In vivo inhibition of hepatitis B virus gene expression by antisense phosphorothioate oligonucleotides. Biochem Biophys Res Commun 218(1):217–223. https://doi.org/10.1006/bbrc.1996.0038. S0006-291X(96)90038-8 [pii]

38. Gura T (1995) Antisense has growing pains. Science 270(5236):575–577

39. Lewis EJ, Agrawal S, Bishop J, Chadwick J, Cristensen ND, Cuthill S, Dunford P, Field AK, Francis J, Gibson V, Greenham AK, Kelly F, Kilkushie R, Kreider JW, Mills JS, Mulqueen M, Roberts NA, Roberts P, Szymkowski DE (2000) Non-specific antiviral activity of antisense molecules targeted to the E1 region of human papillomavirus. Antivir Res 48(3):187–196. S0166354200001297 [pii]

40. Agrawal S, Iyer RP (1997) Perspectives in antisense therapeutics. Pharmacol Therapeut 76:151–160

41. Levin AA (1999) A review of the issues in the pharmacokinetics and toxicology of phosphorothioate antisense oligonucleotides. Biochim Biophys Acta 1489(1):69–84. S0167-4781(99)00140-2 [pii]

42. Galbraith WM, Hobson WC, Giclas PC, Schechter PJ, Agrawal S (1994) Complement activation and hemodynamic changes following intravenous administration of phosphorothioate oligonucleotides in the monkey. Antisense Res Dev 4(3):201–206

43. Black LE, Farrelly JG, Cavagnaro JA, Ahn CH, DeGeorge JJ, Taylor AS, DeFelice AF, Jordan A (1994) Regulatory considerations for oligonucleotide drugs: updated recommendations for pharmacology and toxicology studies. Antisense Res Dev 4(4):299–301

44. Agrawal S (2001) United States Securities and Exchange Commission report. https://www.sec.gov/Archives/edgar/data/861838/000095013501501616/b39654hye8-k.txt. Accessed 9 Dec 2020

45. Bayever E, Iversen PL, Bishop MR, Sharp JG, Tewary HK, Arneson MA, Pirruccello SJ, Ruddon RW, Kessinger A, Zon G et al (1993) Systemic administration of a phosphorothioate oligonucleotide with a sequence complementary to p53 for acute myelogenous leukemia and myelodysplastic syndrome: initial results of a phase I trial. Antisense Res Dev 3(4):383–390

46. de Smet MD, Meenken CJ, van den Horn GJ (1999) Fomivirsen - a phosphorothioate oligonucleotide for the treatment of CMV retinitis. Ocul Immunol Inflamm 7(3–4):189–198

47. Nemunaitis J, Holmlund JT, Kraynak M, Richards D, Bruce J, Ognoskie N, Kwoh TJ, Geary R, Dorr A, Von Hoff D, Eckhardt SG (1999) Phase I evaluation of ISIS 3521, an antisense oligodeoxynucleotide to protein kinase C-alpha, in patients with advanced cancer. J Clin Oncol 17(11):3586–3595. https://doi.org/10.1200/JCO.1999.17.11.3586

48. Grindel JM, Musick TJ, Jiang Z, Roskey A, Agrawal S (1998) Pharmacokinetics and metabolism of an oligodeoxynucleotide phosphorothioate (GEM 91) in cynomologous monkeys following intravenous infusion. Antis Nucl Acid Drug Dev 8:43–52

49. Sereni D, Tubiana R, Lascoux C, Katlama C, Taulera O, Bourque A, Cohen A, Dvorchik B, Martin RR, Tournerie C, Gouyette A, Schechter PJ (1999) Pharmacokinetics and tolerability of intravenous trecovirsen (GEM 91), an antisense phosphorothioate oligonucleotide, in HIV-positive subjects. J Clin Pharmacol 39(1):47–54

50. Agrawal S (1996) Antisense oligonucleotides: towards clinical trials. Trends Biotechnol 14(10):376–387. https://doi.org/10.1016/0167-7799(96)10053-6. 0167-7799(96)10053-6 [pii]

51. Agrawal S, Kandimalla ER (2004) Role of Toll-like receptors in antisense and siRNA [corrected]. Nat Biotechnol 22(12):1533–1537. https://doi.org/10.1038/nbt1042. nbt1042 [pii]

52. Krieg AM, Yi A-K, Matson S, Waldschmidt TJ, Bishop GA, Teasdale R, Koretsky GA, Klinman DM (1995) CpG motife in bacterial DNA trigger direct B-cell activation. Nature 374:546–549

53. Messina JP, Gilkeson GS, Pisetsky DS (1991) Stimulation of in vitro murine lymphocyte

proliferation by bacterial DNA. J Immunol 147:1759–1764

54. Hemmi H, Takeuchi O, Kawai T, Kaisho T, Sato S, Sanjo H, Matsumoto M, Hoshino K, Wagner H, Takeda K, Akira S (2000) A Toll-like receptor recognizes bacterial DNA. Nature 408(6813):740–745. https://doi.org/10.1038/35047123

55. Agrawal S, Martin RR (2003) Was induction of HIV-1 through TLR9? J Immunol 171(4):1621. author reply 1621–1622

56. Agrawal S (1999) Importance of nucleotide sequence and chemical modifications of antisense oligonucleotides. Biochim Biophys Acta 1489:53–68

57. Agrawal S, Kandimalla ER (2000) Antisense therapeutics: is it as simple as complementary base recognition? Mol Med Today 6(2):72–81. S1357-4310(99)01638-X [pii]

58. Eckstein F (1985) Nucleoside phosphorothioates. Annu Rev Biochem 54:367–402

59. Kurpiewski MR, Koziolkiewicz M, Wilk A, Stec WJ, Jen-Jacobson L (1996) Chiral phosphorothioates as probes of protein interactions with individual DNA phosphoryl oxygens: essential interactions of EcoRI endonuclease with the phosphate at pGAATTC. Biochemistry 35(27):8846–8854. https://doi.org/10.1021/bi960261e. bi960261e [pii]

60. Guo M, Yu D, Iyer RP, Agrawal S (1998) Solid-phase stereoselective synthesis of 2′-O-methyl-oligoribonucleoside phosphorothioates using nucleoside bicyclic oxazaphospholidines. Bioorg Med Chem Lett 8:2539–2544

61. Yu D, Kandimalla ER, Roskey A, Zhao Q, Chen L, Chen J, Agrawal S (2000) Stereo-enriched phosphorothioate oligodeoxynucleotides: synthesis, biophysical and biological properties. Bioorg Med Chem 8(1):275–284. S0968-0896(99)00275-8 [pii]

62. Iwamoto N, Butler DCD, Svrzikapa N, Mohapatra S, Zlatev I, Sah DWY, Meena, Standley SM, Lu G, Apponi LH, Frank-Kamenetsky M, Zhang JJ, Vargeese C, Verdine GL (2017) Control of phosphorothioate stereochemistry substantially increases the efficacy of antisense oligonucleotides. Nat Biotechnol 35(9):845–851. https://doi.org/10.1038/nbt.3948. nbt.3948 [pii]

63. Østergaard ME, De Hoyos CL, Wan WB, Shen W, Low A, Berdeja A, Vasquez G, Murray S, Migawa MT, Liang X-H, Swayze EE, Crooke ST, Seth PP (2020) Understanding the effect of controlling phosphorothioate chirality in the DNA gap on the potency and safety of gapmer antisense oligonucleotides. Nucl Acids Res 48:1691. https://doi.org/10.1093/nar/gkaa031

64. Wave Life Sciences Press Release (2019) Wave Life Sciences announces discontinuation of Suvodirsen development for Duchenne Muscular Dystrophy. https://ir.wavelifesciences.com/news-releases/news-release-details/wave-life-sciences-announces-discontinuation-suvodirsen. Accessed 9 Dec 2020

65. Wave Life Sciences Press Release (2019) Wave Life Sciences announces Suvodirsen Phase 1 safety and tolerability data and Phase 2/3 Clinical Trial design. https://ir.wavelifesciences.com/news-releases/news-release-details/wave-life-sciences-announces-suvodirsen-phase-1-safety-and. Accessed 9 Dec 2020

66. Miller PS, Agris CH, Murakami A, Reddy PM, Spitz SA, Ts'o POP (1983) Preparation of oligodeoxyribonucleoside methylphosphonates on a polystyrene support. Nucl Acids Res 11:6225–6241

67. Asai A, Oshima Y, Yamamoto Y, Uochi T, Kusaka H, Akinaga S, Yamashita Y, Pongracz K, Pruzan R, Wunder E, Piatyszek M, Li S, Chin AC, Harley CB, Gryaznov S (2003) A novel telomerase template antagonist (GRN163) as a potential anticancer agent. Cancer Res 63:3931–3939

68. Agrawal S, Goodchild J (1987) Oligodeoxynucleoside methylphosphonates: synthesis and enzymic degradation. Tetrahedron Lett 28:3539–3542

69. Summerton J, Weller D (1993) Uncharged Morpholino-based polymers having phosphorus containing chiral intersubunit linkages

70. Summerton J (1999) Morpholino antisense oligomers: the case for an RNase H-independent structural type. Biochim Biophys Acta 1489:141–158

71. Enterlein S, Warfield KL, Swenson DL, Stein DA, Smith JL, Gamble CS, Kroeker AD, Iversen PL, Bavari S, Mühlberger E (2006) VP35 knockdown inhibits Ebola virus amplification and protects against lethal infection in mice. Antimicrob Agents Chemother 50:984–993

72. Kinali M, Arechavala-Gomeza V, Feng L, Cirak S, Hunt D, Adkin C, Guglieri M, Ashton E, Abbs S, Nihoyanopoulos P, Garraldi EM, Rutherford M, Mcculley C, Popplewell LJ, Graham IR, Dickson G, Wood M, Wells DJ, Wilton SD, Holt T, Kole R, Straub V, Bushby K, Sewry C, Morgan JE, Muntoni F (2009) Restoration of dystrophin expression in Duchenne muscular dystrophy: a single blind placebo-controlled dose

escalation study using morpholino oligomer AVI-4658. Lancet 8:918

73. Aartsma-Rus A, Arechavala-Gomeza V (2018) Why dystrophin quantification is key in the eteplirsen saga. Nat Rev Neurol 14: 454–456

74. Järver P, O'Donovan L, Gait MJ (2014) A chemical view of oligonucleotides for exon skipping and related drug applications. Nucl Acids Ther 24:37–47

75. Egholm M, Buchardt O, Nielsen PE, Berg RH (1992) Peptide Nucleic Acids (PNA). Oligonucleotide analogues with an achiral backbone. J Am Chem Soc 114:1895–1897

76. Rapozzi V, Burm BE, Cogioi S, van der Marel GA, van Boom JH, Quadrifoglio F, Xodo LE (2002) Anti-proliferative effect in chronic myeloid leukaemia cells by antisense peptide nucleic acids. Nucl Acids Res 30:3712–3721

77. Villa R, Folini M, Lualdi S, Veronese S, Daidone MG, Zaffaroni N (2000) Inhibition of telomerase activity by a cell-penetrating peptide nucleic acid construct in human melanoma cells. FEBS Lett 473:241–248

78. Chaubey B, Tripathi S, Ganguly S, Harris D, Casale RA, Pandey VN (2005) A PNA-Transportan conjugate targeted to the TAR region of the HIV-1 genome exhibits both antiviral and virucidal properties. Virology 331:418–428

79. Chaubey B, Tripathi S, Pandey VN (2008) Single acute-dose and repeat-doses toxicity anti-HIV-1 PNATAR-Penetratin conjugates after intraperitoneal administration to mice. Oligonucleotides 18:9–20

80. Good L, Awasthi SK, Dryselius R, Larsson O, Nielsen PE (2001) Bactericidal antisense effects of peptide-PNA conjugates. Nat Biotech 19:360–364

81. Good L, Nielsen PE (1998) Inhibition of translation and bacterial cell growth by peptide nucleic acid targeted to ribosomal RNA. Proc Natl Acad Sci U S A 95:2073–2076

82. Torres AG, Fabani MM, Vigorito E, Williams D, Al-Obaidi N, Wojcechowski F, Hudosn RHE, Seitz O, Gait MJ (2012) Chemical structure requirements and cellular targeting of microRNA-122 by peptide nucleic acids anti-miRs. Nucl Acids Res 40: 2152–2167

83. Inoue H, Hayase Y, Iwai S, Ohtsuka E (1987) Sequence-dependent hydrolysis of RNA using modified oligonucleotide splints and RNase H. Nucl Acids Symp Ser 18:221–224

84. Yu D, Iyer RP, Shaw DR, Lisziewicz J, Li Y, Jiang Z, Roskey A, Agrawal S (1996) Hybrid oligonucleotides: synthesis, biophysical properties, stability studies, and biological activity. Bioorg Med Chem 4(10): 1685–1692. 0968089696001605 [pii]

85. Metelev V, Lisziewicz J, Agrawal S (1994) Study of antisense oligonucleotide phosphorothioates containing segments of oligodeoxynucleotides and 2'-o-methylribonucleotides. Bioorg Med Chem Lett 4:2929–2934

86. Agrawal S, Jiang Z, Zhao Q, Shaw D, Cal Q, Roskey A, Channavajjala L, Saxinger C, Zhang R (1997) Mixed-backbone oligonucleotidesas second generation antisense oligonucleotides: in vitro and in vivo studies. Proc Natl Acad Sci U S A 94:2620–2625

87. Zhou W, Agrawal S (1998) Mixed-backbone oligonucleotides as second-generation antisense agents with reduced phosphorothioate-related side effects. Bioorg Med Chem Lett 8: 3269–3274

88. Sierakowska H, Sambade MJ, Agrawal S, Kole R (1996) Repair of thalassemic human beta-globin mRNA in mammalian cells by antisense oligonucleotides. Proc Natl Acad Sci U S A 93(23):12840–12844

89. Wilton SD, Lloyd F, Carville K, Fletcher S, Honeyman K, Agrawal S, Kole R (1999) Specific removal of the nonsense mutation from the mdx dystrophin mRNA using antisense oligonucleotides. Neuromuscul Disord 9(5): 330–338. S0960896699000103 [pii]

90. FDA Panel (2015) FDA advisory panel votes BioMarin's drisapersentrials not persuasive. https://www.fdanews.com/articles/174238-fda-advisory-panel-votes-biomarins-drisapersentrials-not-persuasive?v=preview. Accessed 9 Dec 2020

91. Aartsma-Rus A (2017) FDA approval of nusinersen for Spinal Muscular Atrophy makes 2016 the year of splice modulating oligonucleotides. Nucl Acids Ther 27:67–69

92. Mercuri E, Darras BT, Chiriboga CA, Day JW, Campbell C, Connolly AM, Iannaccone ST, Kirschner J, Kuntz NL, Saito K, Shieh PB, Tulinius M (2018) Nusinersen versus sham control in later-onset Spinal Muscular Atrophy. New Engl J Med 378:625–635

93. Arzumanov A, Walsh AP, Liu X, Rajwanshi VK, Wengel J, Gait MJ (2001) Oligonucleotide analogue interference with the HIV-1 Tat protein-TAR RNA interaction. Nucleos Nucleot Nucl Acids 20:471–480

94. Lindow M, Kauppinen S (2012) Discovering the first microRNA targeted drug. J Cell Biol 199:407–412

95. Morita K, Hasegawa C, Kaneko M, Tsutsumi S, Sone J, Ishikawa T, Imanishi T, Koizumi M (2001) 2'-O,4'-C-ethylene-

bridged nucleic acids (ENA) with nuclease-resistance and high affinity for RNA. Nucl Acids Res Suppl 1:241–242

96. Lee T, Awano H, Yagi M, Matsumoto M, Watanabe N, Goda R, Koizumi M, Takeshima Y, Matsuo M (2017) 2'-O-methyl RNA/ethylene-bridged nucleic acid chimera antisense oligonucleotides to induce dystrophin exon 45 skipping. Genes 8(2):67. https://doi.org/10.3390/genes8020067. genes8020067 [pii]

97. Renneberg D, Leumann CJ (2002) Watson-Crick base-pairing properties of tricyclo-DNA. J Am Chem Soc 124: 5993–6004

98. Seth PP, Siwkowski A, Allerson CR, Vasquez G, Lee S, Prakash TP, Kinberger G, Migawa MT, Gaus H, Bhat B, Swayze EE (2008) Design, synthesis and evaluation of constrained methoxyethyl (cMOE) and constrained ethyl (cEt) nucleoside analogs. Nucl Acids Symp Ser 52:553–554. https://doi. org/10.1093/nass/nrn280. nrn280 [pii]

99. Flanagan WM, Wagner RW, Grant D, Lin K-Y, Matteucci M (1999) Cellular penetration and antisense activity by a phenoxazine-substituted heptanucleotide. Nat Biotech 17: 48–52

100. Henry S, Stecker K, Brooks D, Monteith D, Conklin B, Bennett CF (2000) Chemically modified oligonucleotides exhibit decreased immune stimulation in mice. J Pharmacol Exp Ther 292(2):468–479

101. Yu D, Wang D, Zhu FG, Bhagat L, Dai M, Kandimalla ER, Agrawal S (2009) Modifications incorporated in CpG motifs of oligodeoxynucleotides lead to antagonist activity of toll-like receptors 7 and 9. J Med Chem 52(16):5108–5114. https://doi.org/10. 1021/jm900730r

102. Zhang R, Iyer RP, Yu D, Tan W, Zhang X, Lu Z, Zhao H, Agrawal S (1996) Pharmacokinetics and tissue disposition of a chimeric oligodeoxynucleoside phosphorothioate in rats after intravenous administration. J Pharmacol Exp Ther 278(2):971–979

103. Kandimalla ER, Temsamani J, Agrawal S (2007) Synthesis and properties of 2'-O-methylribonucleotide methylphosphonate containing chimeric oligonucleotides. Nucleos Nucleot 14:1031–1035

104. Agrawal S, Zhang X, Lu Z, Zhao H, Tamburin JM, Yan J, Cai H, Diasio RB, Habus I, Jiang Z et al (1995) Absorption, tissue distribution and in vivo stability in rats of a hybrid antisense oligonucleotide following oral administration. Biochem Pharmacol 50(4): 571–576. 0006295295001602 [pii]

105. Shen LX, Kandimalla ER, Agrawal S (1998) Impact of mixed-backbone oligonucleotides on target binding affinity and target cleaving specificity and selectivity by Escherichia coli RNase H. Bioorg Med Chem 6:1695–1705

106. Benson MD, Waddington-Cruz M, Berk JL, Polydefkis M, Dyck PJ, Wang AK, Planté-Bordeneuve V, Barroso FA, Merlini G, Obici L, Scheinberg M, Brannagan TH (2018) Inotersen treatment for patients with hereditary transthyretin amyloidosis. New Engl J Med 379:22–31

107. Witztum JL, Gaudet D, Freedman SD, Alexander VA, Digenio A, Williams KR, Yang Q, Hughes SG, Geary RS, Arca M, Stroes ESG, Bergeron J (2019) Volanesorsen and triglyceride levels in Familial Chylomicronemia Syndrome. New Engl J Med 381:531–542

108. Reeskamp LF, Kastelein JJP, Moriarty PM, Duell PB, Catapano AL, Santos RD, Ballantyne CM (2019) Safety and efficacy of mipomersen in patients with heterozygous familial hypercholesterolemia. Atherosclerosis 280: 109–117

109. Frieden M, Orum H (2006) The application of locked nucleic acids in the treatment of cancer. IDrugs 9:706–711

110. Hong D, Kurzrock R, Kim Y, Woessner R, Younes A, Nemunaitis J, Fowler N, Zhou T, Schmidt J, Jo M, Lee SJ, Yamashita M, Hughes SG, Fayad L, Piha-Paul S, Nadella MVP, Mohseni M, Lawson D, Reimer C, Blakey DC, Xiao X, Hsu J, Revenko A, Monia BP, MacLeod AR (2015) AZD9150, a next-generation antisense oligonucleotide inhibitor of STAT3 with early evidence of clinical activity in lymphoma and lung cancer. Sci Transl Med 7:314ra185

111. Seth PP, Yu J, Jazayeri A, Pallan PS, Allerson CR, Østergaard ME, Liu F, Herdewijn P, Egli M, Swayze EE (2012) Synthesis and antisense properties of Fluoro Cyclohexenyl Nucleic Acid (F-CeNA), a nuclease stable mimic of 2'-fluoro RNA. J Org Chem 77: 5074–5085

112. Allart B, Khan K, Rosemeyer H, Schepers G, Hendrix C, Rothenbacher K, Seela F, Van Aerschot A, Herdewijn P (1999) D-Altritol Nucleic Acids (ANA): hybridisation properties, stability, and initial structural analysis. Chem Eur J 5:2424–2431

113. Carroll JB, Warby SC, Southwell AL, Doty CN, Greenlee S, Skotte N, Hung G, Bennett CF, Freier SM, Hayden MR (2011) Potent and selective antisense oligonucleotides targeting single nucleotide polymorphisms in the Huntington Disease gene/allele-specific

silencing of mutant Huntingtin. Mol Ther 19:2178–2185

114. Kamola PJ, Kitson JDA, Turner G, Maratou K, Eriksson S, Panjwani A, Warnock LC, Douillard GA, Moores K, Koppe EL, Wixted WE, Wilson PA, Gooderham NJ, Gany TW, Glark KL, Hughes SA, Edbrooke MR, Parry JD (2015) In silico and in vitro evaluation of exonic and intronic off-target effects form a critical element of therapeutic ASO gamper optimization. Nucl Acids Res 43:8638–8650

115. Kasuya T, Hori S, Watanabe A, Nakajima M, Gahara Y, Rokushima M, Yanagimoto T, Kugimiya A (2016) Ribonuclease H1-dependent hepatotoxicity caused by locked nucleic acid-modified gamper antisense oligonucleotides. Sci Rep 6:30377

116. Burel SA, Hart CE, Cauntey P, Hsiao J, Machemer T, Katz M, Watt A, Bul HH, Younis H, Sabripour M, Freier SM, Hung G, Dan A, Prakash TP, Seth PP, Swayze EE, Bennett CF, Crooke ST, Henry SP (2016) Hepatotoxicity of high affinity gamper antisense oligonucleotides is mediated by RNase H1 dependent promiscuous reduction of very long pre-mRNA transcripts. Nucl Acids Res 44:2093–2109

117. Swayze EE, Siwkowski AM, Wancewicz EV, Migawa MT, Wyrzykiewicz TK, Hung G, Monia BP, Bennett CF (2007) Antisense oligonucleotides containing locked nucleic acid improve potency but cause significant hepatotoxicity in animals. Nucl Acids Res 35:687–700

118. Ferrari N, Bergeron D, Tedeschi A-L, Mangos MM, Paquet L, Renzi PM, Damha MJ (2006) Characterization of antisense oligonucleotides comprising 2′-deoxy-2-′-fluoro-β-d-arabinonucleic acid (FANA). Ann N Y Acad Sci 1082:91–102

119. Egli M, Pallan PS, Allerson CR, Prakash TP, Berdeja A, Yu J, Lee S, Watt A, Gaus H, Bhat B, Swayze EE, Seth PP (2011) Synthesis, improved antisense activity and structural rationale for the divergent RNA affinities of 3′-fluoro hexitol nucleic acid (FHNA and Ara-FHNA) modified oligonucleotides. J Am Chem Soc 133:16642–16649

120. Parmar R, Willoughby JLS, Liu J, Foster DJ, Brighham B, Theile CS, Charisse K, Akinc A, Guidry E, Pei Y, Strapps W, Cancilla M, Stanton MG, Rjaeev KG, Sepp-Lorenzino L, Manoharan M, Meyers R, Maier MA, Jahdav V (2016) 5′-(E)-Vinylphosphonate: a stable phosphate mimic can improve the RNAi activity of siRNA–GalNAc conjugates. ChemBioChem 17:985–989

121. Godfrey C, Desviat LR, Smedsrød B, Piétri-Rouxel F, Denti MA, Disterer P, Lorain S, Nogales-Gadea G, Sardon V, Anwar R, El Andaloussi S, Lehto T, Khoo B, Brolin C, van Roon-Mom WM, Goyenvalle A, Aartsma-Rus A, Arechavala-Gomeza V (2017) Delivery is key: lessons learnt from developing splice-switching antisense therapies. EMBO Mol Med 9:545–557

122. Ray KK, Landmesser U, Leiter LA, Kallend D, Dufour R, Karakas M, Hall T, Troquay RPT, Turner T, Visseren FLJ, Wijngard P, Wright RS (2017) Inclisiran in patients at high cardiovascular risk with elevated LDL cholesterol. New Engl J Med 376:1430–1440

123. Godhino BMDC, Coles AH, Khvorova A (2019) Conjugate-mediated delivery of RNAi-based therapeutics: enhancing pharmacokinetics-pharmacodynamics relationships of medicinal oligonucleotides. In: Agrawal S, Gait MJ (eds) Advances in nucleic acid therapeutics. Royal Society of Chemistry, London, pp 206–232

124. Killanthottathil GR, Manoharan M (2019) Liver-targeted RNAi therapeutics: principles and applications. In: Agrawal S, Gait MJ (eds) Advances in nucleic acid therapeutics. Royal Society of Chemistry, London, pp 233–265

125. Chirigos MA, Papademetriou V, Bartocci A, Read E, Levy HB (1981) Immune response modifying activity in mice of polyinosinic:polycytidylic acid stabilized with poly-L-lysine, in carboxymethylcellulose [Poly-ICLC]. Int J Immunopharmacol 3:329–337

126. Pisetsky DS (1996) Immune activation by bacterial DNA: a new genetic code. Immunity 5:303–310

127. Boller T, Felix G (2009) A renaissance of elicitors: perception of microbe-associated molecular patterns and danger signals by pattern-recognition receptors. Annu Rev Plant Biol 60:379–406

128. Palm NW, Medzhitov R (2008) Pattern recognition receptors and control of adaptive immunity. Immunol Rev 227:221–233

129. Blasius AL, Beutler B (2011) Intracellular toll-like receptors. Immunity 32:305–315

130. Kandimalla ER, Agrawal S (2005) Agonists of toll-like receptor 9. In: Toll and toll-like receptors: an immunologic perspective. Molecular Biology Intelligence Unit. Springer, Boston, MA, pp 181–212

131. Bauer S, Kirschning CJ, Häcker H, Redecke V, Hausmann S, Akira S, Wagner H, Lipford GB (2001) Human

TLR9 confers responsiveness to bacterial DNA via species-specific CpG motif recognition. Proc Natl Acad Sci U S A 98:9237–9242

132. Cho WG, Albuquerque RJC, Kleinman ME, Taralio V, Greco A, Nozaki M, Green MG, Baffi JZ, Ambati BK, De Falco M, Alexander JS, Brunetti A, De Falco S, Anbelti J (2009) Small interfering RNA-induced TLR3 activation inhibits blood and lymphatic vessel growth. Proc Natl Acad Sci U S A 106:7137–7142

133. Kleinman ME, Yamada K, Takeda A, Chandrasekaran V, Nozaki M, Baffi JZ, Albuquerque RJC, Yamasaki S, Itaya M, Pan Y, Appukuttan B, Gibbs D, Yang Z, Kariko K, Ambati BK, Eilgus TA, DiPietro LA, Sakurai E, Zhang K, Smith JR, Taylor EW, Ambati J (2008) Sequence- and target-independent angiogenesis suppression by siRNA via TLR3. Nature 452:591–597

134. Lan T, Wang D, Bhagat L, Philbin VJ, Yu D, Tang JX, Putta MR, Sullivan T, La Monica N, Kandimalla ER, Agrawal S (2013) Design of synthetic oligoribonucleotide-based agonists of Toll-like receptor 3 and their immune response profiles in vitro and in vivo. Org Biomol Chem 11:1049–1058

135. Putta MR, Zhu FG, Wang D, Bhagat L, Kandimalla ER, Agrawal S (2010) Peptide conjugation at the 5′-end of oligodeoxynucleotides abrogates Toll-Like Receptor 9-mediated immune stimulatory activity. Bioconjug Chem 21:39–45

136. Agrawal S, Kandimalla ER (2001) Antisense and/or immunostimulatory oligonucleotide therapeutics. Curr Cancer Drug Targets 1(3):197–209

137. Jackson S, Lenting J, Kopp J, Murray L, Ellison W, Rhee M, Shockey G, Akelia L, Ery K, Hayward WL, Janssen RS (2018) Immunogenicity of a two-dose investigational hepatitis B vaccine, HBsAg-1018, using a toll-like receptor 9 agonist adjuvant compared with a licensed hepatitis B vaccine in adults. Vaccine 36:668–674

138. Guyadar D, Bogomolv P, Kobalava Z, Moiseev V, Szlavik J, Astruc B, Varkonyi L, Sullivan T (2011) 1209 IMO-2025 plus Ribavirin gives substantial first-dose viral load reductions, cumulative anti-viral effect, is well tolerated in naive genotype HCV patients: a Phase 1 trial. J Hepatol 54(Supp 1):S478

139. Jiang W, Zhu FG, Bhagat L, Yu D, Tang JX, Kandimalla ER, La Monica N, Agrawal S (2013) A toll-like receptor 7, 8, and 9 antagonist inhibits Th1 and Th17 responses and inflammasome activation in a model of

IL-23-induced psoriasis. J Investig Dermatol 133:1777–1784

140. El-Andaloussi S, Johansson HJ, Holm T, Langel U (2007) A novel cell-penetrating peptide, M918, for efficient delivery of proteins and peptide nucleic acids. Mol Ther 15:1820–1826

141. Poeck H, Besch R, Hartmann G (2008) 5′-triphosphate-siRNA: turning gene silencing and Rig-I activation against melanoma. Nat Med 14:1256–1263

142. Case CL (2011) Regulating caspase-1 during infection: roles of NLRs, AIM2, and ASC. Yele J Biol Med 84:333–343

143. Arnold AE, Malek-Adamian E, Le PU, Meng A, Martinez-Montero S, Petrecca K, Damha MJ, Sholchet MS (2018) Antibody-antisense oligonucleotide conjugate downregulates a key gene in glioblastoma stem cells. Mol Ther Nucl Acids 11:518–527

144. Cuellar TL, Barnes D, Nelson C, Tanguay J, You S-F, Wen X, Scales SJ, Gesch J, Davis D, van Brabant SA, Leake D, Vandlen R, Sieber CW (2015) Systematic evaluation of antibody-mediated siRNA delivery using an industrial platform of THIOMAB–siRNA conjugates. Nucl Acids Res 43:1189–1203

145. Langel U (2015) Cell-penetrating peptides. In: Methods and protocols. Methods in molecular biology, 2nd edn. Springer, New York, NY

146. Jearawiriyapaisarn N, Moulton HM, Buckley B, Roberts J, Sazani P, Fucharoen S, Iversen PL, Kole R (2008) Sustained dystrophin expression induced by peptide-conjugated morpholino oligomers in the muscles of mdx mice. Mol Ther 16:1624–1629

147. Passini MA, Hanson GJ (2018) Exon skipping oligomer conjugates for muscular dystrophy

148. Betts C, Saleh AF, Arzumanov AA, Hammond SM, Godfrey C, Coursindel T, Gait MJ, Wood MJA (2012) A new generation of peptide-oligonucleotide conjugates with improved cardiac exon skipping activity for Duchenne muscular dystrophy treatment. Mol Ther Nucl Acids 1:e38

149. Klein AF, Varela M, Arandel L, Holland A, Naouar N, Arzumanov A, Seoane D, Revillod L, Bassez G, Ferry A, Jauvin D, Gourdon G, Puymirat J, Gait MJ, Furling D, Wood MJ (2019) Peptide-conjugated oligonucleotides evoke long-lasting myotonic dystrophy correction in patient-derived cells and mice. J Clin Investig 129:4739–4744

150. Juliano RL (2016) The delivery of therapeutic oligonucleotides. Nucl Acids Res 14: 6518–6548

151. Bacsa B, Horváti K, Bosze S, Andreae F, Kappe CO (2008) Solid-phase synthesis of difficult peptide sequences at elevated temperatures: a critical comparison of microwave and conventional heating technologies. J Org Chem 73:7532–7542

152. Shubham S, Lin L-H, Udofot O, Krupse S, Giangrande PH (2019) Prostate-specific membrane antigen (PMSA) aptamers for prostate cancer imaging and therapy. In: Agrawal S, Gait MJ (eds) Advances in nucleic acid therapeutics. Royal Society of Chemistry, London, pp 339–366

153. Zon G (2019) Aptamers and clinical applications. In: Agrawal S, Gait MJ (eds) Advances in nucleic acids therapeutics. Royal Society of Chemistry, London, pp 367–399

154. Strenkowska M, Grzela R, Majewski M, Wnek K, Kowalski J, Lukaszewicz M, Zuberek J, Darzynkiewicz E, Kuhn AN, Sahin U, Jemielty J (2016) Cap analogs modified with 1,2-dithiodiphosphate moiety protect mRNA from decapping and enhance its translational potential. Nucl Acids Res 44: 9578–9590

155. Andries O, McCafferty S, De Smedt SC, Weiss R, Sanders NN, Kitada T (2015) N (1)-methylpseudouridine-incorporated mRNA outperforms pseudouridine-incorporated mRNA by providing enhanced protein expression and reduced immunogenicity in mammalian cell lines and mice. J Control Release 217:337–344

156. Svitkin YV, Cheng YM, Chakraborty T, Presnyak V, John M, Sonenberg N (2017) N1-methyl-pseudouridine in mRNA enhances translation through eIF2-α-dependent and independent mechanisms by increasing ribosome density. Nucl Acids Res 45:6023–6036

157. Oberli MA, Rechmuth AM, Dorkin JR, Mitchell MJ, Fenton OS, Jaklenec A, Anderson DG, Langer R, Blankschtein D (2017) Lipid nanoparticle assisted mRNA delivery for potent cancer immunotherapy. Nano Lett 17:1326–1335

158. Yin H, Song C-Q, Suresh S, Wu Q, Walsh S, Rhym LH, Mintzer E, Bolukbasi MF, Zhu LJ, Kauffman K, Mou H, Ovberholzer A, Ding J, Kwan S-Y, Bogorad RL, Zatsepin TS, Koteliansky V, Wolfe SA, Xue W, Langer R, Anderson DG (2017) Structure-guided chemical modification of guide RNA enables potent non-viral in vivo genome editing. Nat Biotech 35:1179–1187

159. Hendel A, Bak RO, Clark JT, Kennedy AB, Ryan DE, Roy S, Steinfeld I, Lunstad BD, Kaiser RJ, Wilkens AB, Bacchette R, Tsalenko A, Dellinger D, Bruhn L, Porteus MH (2015) Chemically modified guide RNAs enhance CRISPR-Cas genome editing in human primary cells. Nat Biotech 33: 985–989

160. Cromwell CR, Sung K, Park J, Krysler AR, Jovel J, Kim SK, Hubbard BP (2018) Incorporation of bridged nucleic acids into CRISPR RNAs improves Cas9 endonuclease specificity. Nat Commun 9:1448

161. Geny S, Hosseini ES, Concordet J-P, Giovannangeli C (2019) CRISPR-based technologies for genome engineering: properties, current improvements and applications in medicine. In: Agrawal S, Gait MJ (eds) Advances in nucleic acid therapeutics. Royal Society of Chemistry, London, pp 400–433

2

Conjugation of Nucleic Acids and Drugs to Gold Nanoparticles

Paula Milán-Rois, Ciro Rodriguez-Diaz, Milagros Castellanos and Álvaro Somoza ⓘ

Abstract

Gold nanoparticles (AuNPs) can be used as carriers for biomolecules or drugs in cell culture and animal models. Particularly, AuNPs ease their internalization into the cell and prevent their degradation. In addition, engineered AuNPs can be employed as sensors of a variety of biomarkers, where the electronic and optical properties of the AuNPs are exploited for a convenient, easy, and fast read out. However, in all these applications, a key step requires the conjugation of the different molecules to the nanoparticles. The most common approach exploits the great affinity of sulfur for gold. Herein, we summarize the methods used by our group for the conjugation of different molecules with AuNPs. The procedure is easy and takes around 2 days, where the reagents are slowly added, following an incubation at room temperature to ensure the complete conjugation. Finally, the unbound material is removed by centrifugation.

Key words Gold nanoparticles, Spherical nucleic acid, Functionalization, Oligonucleotides, Nanomedicine, Metal nanoparticles, Conjugation, Drug delivery, Sensors

1 Introduction

Oligonucleotides and drugs face some challenges for their optimal delivery in cells and animal models. Particularly, oligonucleotides (e.g., antisense, gapmers, and siRNAs) usually present low stability and suffer from reduced cell internalization and selectivity [1, 2] and, for these reasons, transfection reagents such as lipofectamine are usually employed to improve delivery. On the other hand, drugs can be too hydrophobic and require solubilizing molecules (e.g., dimethylsulfoxide [DMSO], ethanol). However, these kinds of chemicals present critical restrictions such as cytotoxicity or limited loading. To overcome these drawbacks, delivery systems based on nanoparticles can be employed [3]. There are different types of nanoparticles such as liposomes, micelles, dendrimers, inorganic particles, carbon-based nanostructures, viral nanocarriers, polymeric, peptide or metallic nanoparticles, etc. [4–12]. Each vehicle presents different characteristics that can be exploited to address specific challenges related to the delivery of bioactive molecules.

Among the different systems, gold nanoparticles (AuNPs) present excellent properties for the delivery of oligonucleotides because of their low toxicity, cost, and particularly their ease of preparation and functionalization [13]. AuNPs can be synthesized in the laboratory through simple methods, such as the one described by Turkevich [14] and detailed in Subheading 3.1.

The properties of AuNPs can be tuned through their modification with oligonucleotides. When the nanoparticles are densely loaded with oligonucleotides, the resulting nanostructures are known as spherical nucleic acids (SNA) [15]. This kind of nanostructure presents interesting features, such as high internalization in a wide variety of cells and low toxicity. Therefore, these derivatives can be employed for multiple applications, such as drug delivery systems, gene therapy and regulation, or molecular diagnosis [16, 17].

Regarding the vehiculization of therapeutics, AuNPs can be used for the delivery of hydrophobic drugs such as paclitaxel, doxorubicin, or AZD8055 without affecting their effectiveness [18, 19]. On the other hand, AuNPs functionalized with oligonucleotides (e.g., siRNAs, gapmers) could be used as a substitute for transfection reagents in different applications involving gene regulation, or even immunomodulatory processes, for the treatment of diseases such as cancer, sepsis, skin disorders, diabetes, etc. [16, 20–22].

In the case of diagnostics, it is worth mentioning that fast and accurate point-of-care diagnostic systems are critical in personalized medicine. In particular, nucleic acid detection is of great importance for the diagnosis and treatment of many diseases caused by genetic mutations, infectious agents, or other physiologically abnormal circumstances. Conventional methods such as RT-PCR offer high accuracy and sensitivity; however, these methods are not suitable for routine diagnosis because they are time-consuming and need highly trained personnel and expensive equipment. One development that seems to simplify the nucleic acid detection and we study in the lab is the use of SNA based on a single-stranded oligonucleotide with a unique stem-loop structure (Molecular Beacon, MB) [23, 24].

This chapter describes how to conjugate drugs or oligonucleotides to AuNPs, which can be further used as delivery systems of therapeutics and sensors.

To attach any compound to AuNPs, the high affinity of thiol groups to gold could be exploited. Thus, the molecules (e.g., oligonucleotides, drugs) should be functionalized with linkers containing sulfur-based moieties, such as thiols or dithiolanes [25], which are commented in this chapter.

AuNPs conjugation requires a few simple steps of addition, incubation, and washes. The method might change slightly depending on the linker employed for the conjugation, which can be designed to control the release or stability of the cargo. In general, the use of dithiolane provides more robust structures and can be achieved in few hours, whereas the use of thiols implies more than 1 day. For the reader's convenience, we have included the preparation of the dithiolane linkers used in our group. The approach can be used for the conjugation of drugs, polymers (e.g., polyethylene glycol [PEG]), or the preparation of oligonucleotides in a DNA synthesizer using a tailored solid support, usually based on controlled pore glass (CPG).

2 Materials

2.1 AuNP Synthesis

2.1.1 Materials

- 250 mL round-bottom glass flask.
- Septum for a 250 mL round-bottom flask.
- 3.5 cm long magnet.
- 0.3 μm fritted filter for vacuum filtration.
- 250 mL Erlenmeyer flask with an output for vacuum.
- Plastic material such as conical centrifuge tubes and microcentrifuge tubes.
- 1 mL quartz cuvette.

2.1.2 Reagents

- Gold solution: 945.2 μM Hydrogen tetrachloroaurate (III) hydrate ($AuCl_4H_3O$) in 100 mL autoclaved Milli-Q grade water.
- Ultrapure reagent-grade water.
- Sodium citrate solution: 40 mM sodium citrate tribasic dihydrate (118 mg) in 10 mL autoclaved Milli-Q grade water.

2.1.3 Equipment

- Hot plate (7 cm radius) with magnetic stirring.
- Reflux column.
- Vacuum pump.
- UV-Vis spectrophotometer.

2.2 Dithiolane-Based Linkers Synthesis

2.2.1 Materials

– 50 mL round-bottom flask.
– 2-cm long magnet.
– 1 septum.
– Thin-layer chromatography (TLC) (sheets of silica gel 60F254).
– Filter paper.
– 2000 KDa tubing membrane.

2.2.2 Reagents

– Lipoic acid.
– *N*-hydroxysuccinimide (NHS).
– Tetrahydrofuran (THF).
– *N,N'*-dicyclohexylcarbodiimide (DCC).
– Ethyl acetate (AcOEt).
– Methoxypolyethylene glycol amine (PEG-NH_2).
– Drug with a primary amine (e.g., Gemcitabine).
– Dimethylformamide (DMF).
– Dichloromethane (CH_2Cl_2).
– Methanol (MeOH).
– Threoninol.
– 4,4'-Dimethoxytrityl chloride (DMTrCl).
– Dry pyridine (Py).
– Hexane.
– Succinic anhydride.
– Dry CH_2Cl_2.
– 4-(Dimethylamino)pyridine (DMAP).
– *N,N*-Diisopropylethylamine (DIPEA).
– Distilled water.
– Magnesium sulfate anhydrous ($MgSO_4$).
– 1-Hydroxybenzotriazole (HBO^t).
– Acetonitrile (MeCN).
– CPG: Aminopropyl-CPG, 1000 Å.
– Caping reagent A (CAP A): THF/pyridine/acetic anhydride (8:1:1).
– Caping reagent B (CAP B): 10% Methylimidazole in THF.

2.2.3 Equipment

– Flash column chromatography using silica gel (60 Å, 230 × 400 mesh).
– Rotavapor.

2.3 Functionalization of AuNPs

– Plastic material: one microcentrifuge tube per condition.

– A 96-well plate for absorbance measurements in a plate reader.

2.3.1 Materials

2.3.2 Reagents

– Oligonucleotides with sulfide-based modifications at micromolar concentration.

– Annealing buffer 3×: 30 mM Tris-HCl, 3 mM EDTA, 150 mM NaCl.

– Tris (2-carboxyethyl) phosphine hydrochloride solution (TCEP).

– Gold nanoparticles (AuNP) with a diameter of 13 ± 2 nm (*see* Subheading 3.1).

– Sodium chloride solution: 5 M NaCl.

– Oligonucleotide quantification kit (e.g., Quant-iT™ OliGreen™ ssDNA Assay Kit, Qubit™ ssDNA Assay Kit), including the Quant-iT OliGreen® ss DNA Reagent, TE 20× buffer, and oligonucleotide standard.

– Autoclaved Milli-Q grade water.

– PEG modified with a dithiolane group (*see* Subheading 3.2.2).

– Chemotherapeutic drugs with dithiolane-based linker (e.g., gemcitabine) (*see* Subheading 3.2.3).

2.3.3 Equipment

– Benchtop centrifuge.

– Vortex mixer.

– Orbital shaker.

– Plate reader suitable for absorbance and fluorescence determinations using 96-well plates.

– Evaporating centrifuge.

3 Methods

3.1 AuNP Synthesis

For the preparation of AuNP (13 ± 2 nm) the Turkevich method is used [14] as follows:

1. Turn on the heating plate to 140 °C.

2. Add 100 μL of $HAuCl_4$ solution (94.52 μmol) in a 250 mL round bottom flask containing 100 mL of sterile water.

3. Add the magnet to the round bottom flask.

4. Place the round bottom flask in the heating plate while stirring at 700 rpm approximately, with a reflux system, and heat it to reflux.

5. Prepare the sodium citrate tribasic dihydrate solution in a 50 mL conical centrifuge tube.

6. When the mixture boils, add the citrate solution quickly while stirring at 700 rpm.

7. Wait for 15 min and then remove the round bottom flask from the heating plate. During this period, the color of the solution turns from yellow to red (*see* **Note 1**).

8. Leave the mixture stirring at 300 rpm at room temperature and protect from the light for 16 h.

9. Filter the solution using a 0.3 μm fritted filter with the help of vacuum.

10. Determine nanoparticles' size: for proper characterization of gold nanoparticles, the size should be measured by TEM and the concentration determined by UV-Vis spectrophotometry using the Beer-Lambert law. It requires measuring the UV-Vis absorbance at 520 nm and using the corresponding extinction coefficient (ε) for its size [26]. For instance, to determine the concentration of 13 ± 1 nm AuNPs, which have an extinction coefficient (ε) of 2.7×10^8 M^{-1} cm^{-1}, you should use the following equation (Eq. 1):

$$\text{concentration} = \frac{A}{\varepsilon \times l} \tag{1}$$

where A is the absorbance at 520 nm, l is the optical path in cm, and ε extinction coefficient in M^{-11} cm^{-1}.

3.2 Dithiolane-Based Linkers Synthesis

The preparation of the dithiolane-based derivatives of drugs (**3**) and PEG (**2**) is summarized in Fig. 1 and described in the following instructions. In the case of oligonucleotides, the required solid support (CPG) containing a dithiolane moiety for the preparation of oligonucleotides is also described (**7**).

3.2.1 Compound 1 [2,5-Dioxopyrrolidin-1-yl(R)-5-(1,2-Dithiolan-3-yl) Pentanoate]

1. Dissolve lipoic acid (1 equiv.) and *N*-hydroxysuccinimide (1.2 equiv.) in tetrahydrofuran (0.5 M).

2. Stir the solution at 0 °C for 10 min.

3. Dissolve *N*,*N*′-dicyclohexylcarbodiimide (1.2 equiv.) in tetrahydrofuran (3.5 M) and add it slowly to the lipoic acid and *N*-hydroxysuccinimide solution obtained in **step 1**.

4. Stir the reaction at room temperature for 5 h.

5. Filter the mixture using a filter paper and wash the solid with cold ethyl acetate. Evaporate the solvent under vacuum to obtain compound **1** as a yellow oil.

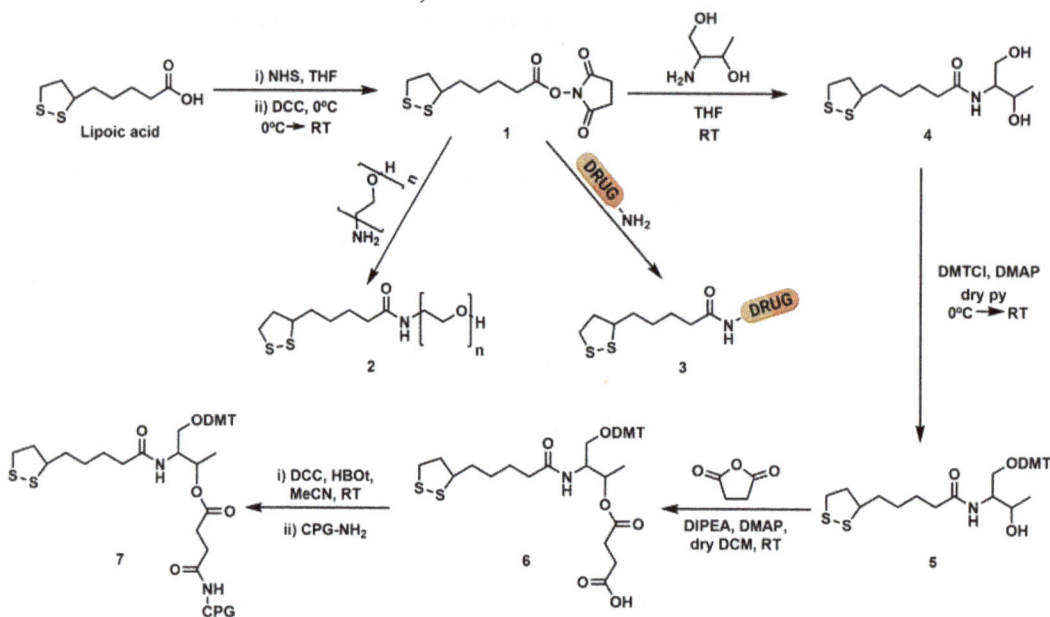

Fig. 1 Schematic representation of the synthesis of dithiolane-modified products: PEG, drug, and CPG

3.2.2 PEG (2)

1. Dissolve compound **1** (2 equiv.) and PEG-NH$_2$ (1 equiv.) in tetrahydrofuran.
2. Stir the reaction at room temperature for 18 h.
3. Purify the crude by dialysis with a 2000 KDa tubing membrane.
4. Stir the solution for 18 h to obtain compound **2**.

3.2.3 Drug-Modified Linker (3)

1. Dissolve compound **1** (2 equiv.) and a drug containing a primary amine (e.g., Gemcitabine) (1 equiv.) in dimethylformamide (0.1 M).
2. Stir the reaction at room temperature for 24 h.
3. Eliminate the solvent in vacuum.
4. Purify the crude by flash chromatography (CH$_2$Cl$_2$:MeOH/ 25:1) to obtain compound **3**.

3.2.4 Compound 4: N-(1,3-Dihydroxybutan-2-yl)-5-(1,2-Dithiolan-3-yl) Pentanamide

1. Dissolve compound **1** (1 equiv.) and threoninol (1.1 equiv.) in THF (0.15 M).
2. Stir the solution at room temperature for 18 h.
3. Eliminate the solvent in vacuum.
4. Purify the crude by flash chromatography (CH$_2$Cl$_2$:MeOH/ 25:1) to obtain compound **4** as a yellow oil.

3.2.5 Compound 5: N-(1-(bis(4-Methoxyphenyl)(Phenyl)Methoxy)-3-Hydroxybutan-2-yl)-5-(1,2-Dithiolan-3-yl)Pentanamide

1. Dissolve compound **4** (1 equiv.) in dry pyridine (0.3 M).
2. Dissolve DMTrCl (1.2 equiv.) in dry pyridine (1 M) and add to the compound **4** solution in dry pyridine at 0 °C.
3. Stir the reaction at 0 °C for 30 min, then at room temperature for 18 h.
4. Eliminate the solvent in vacuum.
5. Purify the crude by flash chromatography (Hexane:AcOEt/1:1) to obtain compound **5** as a beige foam.

3.2.6 Compound 6: 4-((3-(5-(1,2-Dithiolan-3-yl)Pentanamido)-4-(bis(4-Methoxyphenyl)(Phenyl)Methoxy)Butan-2-yl)oxy)-4-Oxobutanoic Acid

1. Dissolve compound **5** (1 equiv.), DMAP (0.1 equiv.) and DIPEA (1.4 equiv.) in dry CH_2Cl_2 (0.13 M).
2. Dissolve succinic anhydride (1.3 equiv.) in dry CH_2Cl_2 (0.3 M) and add the solution slowly to the mixture prepared in the previous step at 0 °C.
3. Stir the reaction at room temperature for 18 h.
4. Wash the solution with water 3 times.
5. Dry the organic layer with $MgSO_4$.
6. Eliminate the solvent in vacuum to obtain compound **6**.

3.2.7 Compound 7: 4-((3-(5-(1,2-Dithiolan-3-yl)Pentanamido)-4-(bis(4-Methoxyphenyl)(Phenyl)Methoxy)Butan-2-yl)oxy)-4-Oxobutanamide CPG

1. Dissolve compound **6** (1 equiv.) in MeCN (0.05 M).
2. Dissolve DCC (1 equiv.) and HBO^t (1 equiv.) in MeCN (0.2 M) and add to compound **6** solution at room temperature for 3 h.
3. Filter the solution with filter paper and add to the CPG (5 equiv. in mg).
4. Stir the mixture for 3 h at room temperature.
5. Remove the solvent and wash the CPG with MeOH three times and with CH_2Cl_2 three times.
6. Dry the CPG.
7. Add a mixture of capping reagents [CAP A:CAP B (1:1)] (1 mL per 175 mg of CPG).
8. Stir the solution for 1 h at RT.
9. Remove the solvent and wash the CPG with MeOH three times and with MeCN three times.

3.3 AuNP Functionalization with Thiol-Modified Oligonucleotides

Oligonucleotides can be easily attached to AuNPs using a thiol-based linker, which is commercially available, and most oligonucleotide providers offer this modification. However, the thiol group should be deprotected, as detailed below, before incubating the oligonucleotides with AuNPs (Fig. 2).

1. Incubate the oligonucleotide with TCEP (*see* **Note 2**) using a 100× excess relative to the oligonucleotide's thiol (*see* **Note 3**)

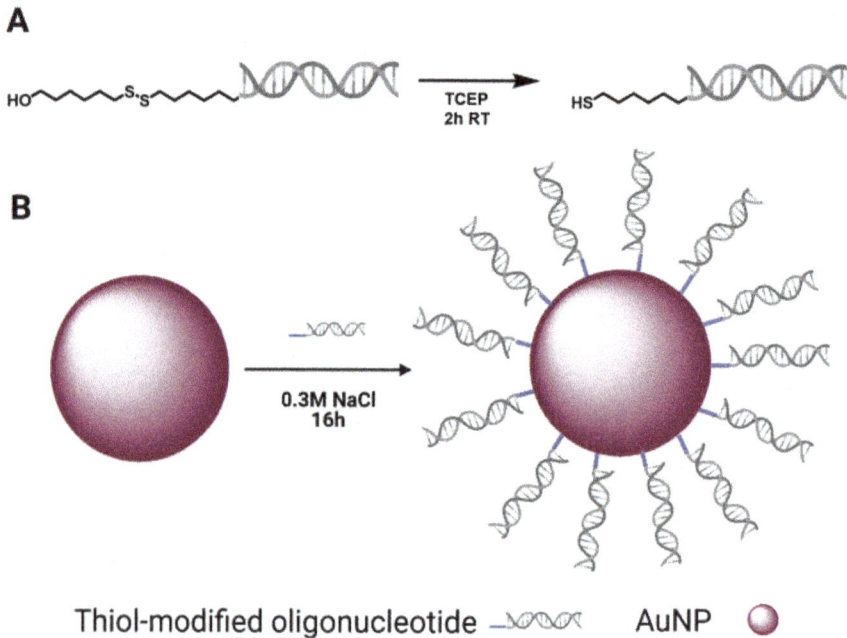

Fig. 2 Schematic representation of: (**a**) Deprotection of oligonucleotides bearing a thiol moiety. (**b**) Functionalization of AuNPs with thiol-modified oligonucleotides

for 2 h at room temperature and moderate agitation on a mini-shaker (e.g., for deprotect 500 µL of an oligonucleotide solution at 20 µM (i.e., 20 pmol/µL) use 2 µL of TCEP at 0.5 M) (*see* **Note 4**).

2. Add the deprotected oligonucleotide slowly to the AuNP solution prepared at 12 nM.

3. Incubate the mixture for 45 min at room temperature and moderate agitation.

4. Add 60 µL NaCl solution to a final concentration of 0.3 M through the addition of small volumes (e.g., 5–10 µL) (*see* **Note 5**). Vortex the solution quickly after each addition of NaCl solution and incubate the sample for at least 10 min on a mini-shaker between each addition.

5. Incubate the sample for 16 h at room temperature on a mini-shaker with moderate agitation.

6. Remove any unbound material by centrifugation at 13.2 rpm and 4 °C for 30 min. After the centrifugation, remove the supernatant and save it for later use. The pellet should be resuspended by vortexing in sterile water using the same volume of water removed to keep the concentration constant. Repeat the cleaning **step 3** times (*see* **Note 6**).

7. Evaporate the collected supernatants and use the Quant-iT™ OliGreen™ ssDNA Assay Kit protocol to determine the

unbound oligonucleotide from the solution (*see* Subheading 3.4) (*see* **Note 7**).

3.4 Oligonucleotide Quantification

For the quantification of oligonucleotides in the supernatant, use an oligonucleotide quantification kit. In this case, the Quant-iT OliGreen kit is used. The general procedure is as follows.

3.4.1 Standard Curve Preparation

1. Prepare a standard curve for each oligonucleotide using at least 5 dilutions of the specific oligonucleotide in TE buffer (1×), e.g., 0, 1000, 750, 500, 250 ng/mL.

2. Prepare a solution of Quant-iT OliGreen® ss DNA Reagent (2 μg/mL) in TE (1×).

3. Mix each oligonucleotide dilution from **step 1** with 1 mL Oligreen solution prepared in **step 2**. Incubate the solution for 5 min at room temperature protected from light.

4. Take 200 μL of each solution prepared in **step 3** and measure it in a plate reader (excitation 480 nm, emission 520 nm).

5. Plot the data in a concentration (ng/mL, *X*-axis) vs. absorbance (a.u., *Y*-axis) graph. Fit the data to a simple linear regression model and use this equation to calculate future concentrations.

3.4.2 Oligonucleotide Quantification in AuNPs

1. Evaporate to dryness the supernatant collected during the cleaning of modified AuNPs (*see* **step 7** in Subheading 3.3).

2. Resuspend the pellet in 1 mL TE (1×).

3. Prepare a solution of Oligreen reagent (2 μg/mL) in TE (1×).

4. Mix the resuspended supernatant with 1 mL Oligreen solution (*see* **step 2** in Subheading 3.4.1). Incubate the solution for 5 min at room temperature protected from light.

5. Take 200 μL of each solution prepared in **step 4** and measure it in a plate reader (excitation 480 nm, emission 520 nm).

6. Interpolate the data obtained in the standard curve equation (*see* **step 5** in Subheading 3.4.1) to determine the amount of unbounded oligonucleotide.

3.5 AuNP Functionalization with Dithiolane-Modified Oligonucleotides or Drugs

Oligonucleotides could be attached to AuNPs in a faster way using a dithiolane-based linker, which does not require a deprotection step, as in the case of thiols (Fig. 3).

1. Add the oligonucleotide to 1 mL 12 nM of 13 ± 2 nm gold nanoparticles (AuNP) (*see* **Note 3**).

3.5.1 AuNP Functionalization with Dithiolane-Modified Oligonucleotides

2. Incubate the solution for 15 min at room temperature on a mini-shaker at a moderate speed.

3. Add 60 μL NaCl solution to a final concentration of 0.3 M through the addition of small volumes (e.g., 5–10 μL) (*see*

Note 5). Vortex the solution quickly after each addition of NaCl solution and incubate the sample for at least 10 min on a mini-shaker between each addition.

4. Incubate the sample for 4 h at room temperature on a mini-shaker at a moderate speed.

5. Continue with the washing steps, as described previously (*see* **step 6** in Subheading 3.3) and the quantification of the unbound material (*see* Subheading 3.4) (*see* **Notes 6** and **7**).

3.5.2 AuNP Functionalization with Dithiolane-Modified Drugs

In this case, the drugs have to be modified with a dithiolane-based linker (Fig. 4), and the AuNPs should be stabilized with oligonucleotides or PEG containing a sulfide-based linker. In this case, for 1 mL of 13 ± 2 nm AuNP (12 nM) add at least 2000 pmol of stabilizing agent (e.g., PEG, oligonucleotide) and then the required amount of the modified drug for a total of 10,000 pmol (stabilizing agent + drug) in the solution.

1. Add the stabilizing agent and incubate it for 15 min at room temperature on a mini-shaker at a moderate speed.

2. Add the modified drug very slowly and incubate it for 16 h at room temperature on a mini-shaker at a moderate speed (*see* **Note 5**).

3. Wash the nanoparticles as described in **step 6** in Subheading 3.3 (*see* **Note 6**).

4. Evaporate the collected supernatant to the initial volume (1 mL) to quantify the drug and determine the nanoparticles' loading.

5. Measure the supernatant in a spectrophotometer (according to the specific absorbance of the drug) and calculate the unbound drug using the Beer-Lambert formula (*see* **Note 8**).

4 Notes

1. AuNP solution should be kept in darkness.

2. Keep TCEP under an inert atmosphere to prevent its oxidation. Once opened, store the compound in 20 µL aliquots at $-20\ °C$.

3. Duplexes should be annealed from their corresponding oligonucleotides before conjugation. In short, combine equal concentration and volume of each strand and add the same volume of annealing buffer (3×). Then, incubate the sample at 95 °C for 10 min and leave to cool slowly to room temperature.

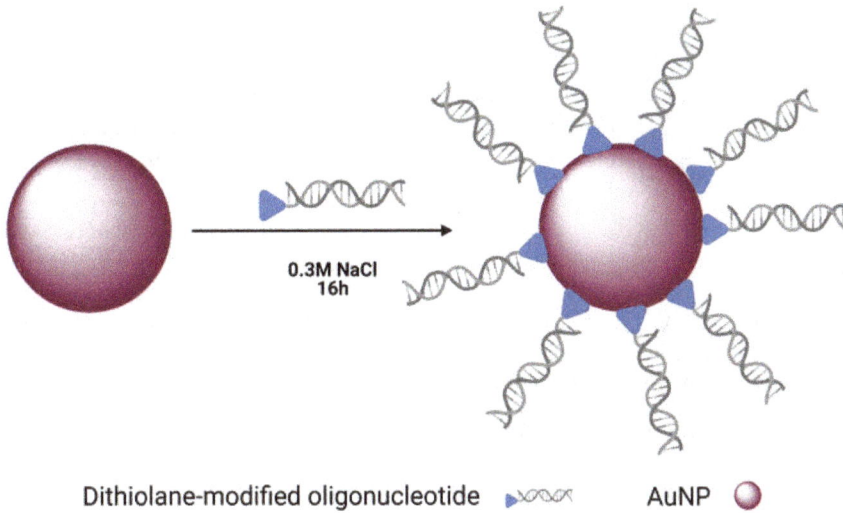

Fig. 3 Schematic representation of AuNPs functionalization with dithiolane-modified oligonucleotides

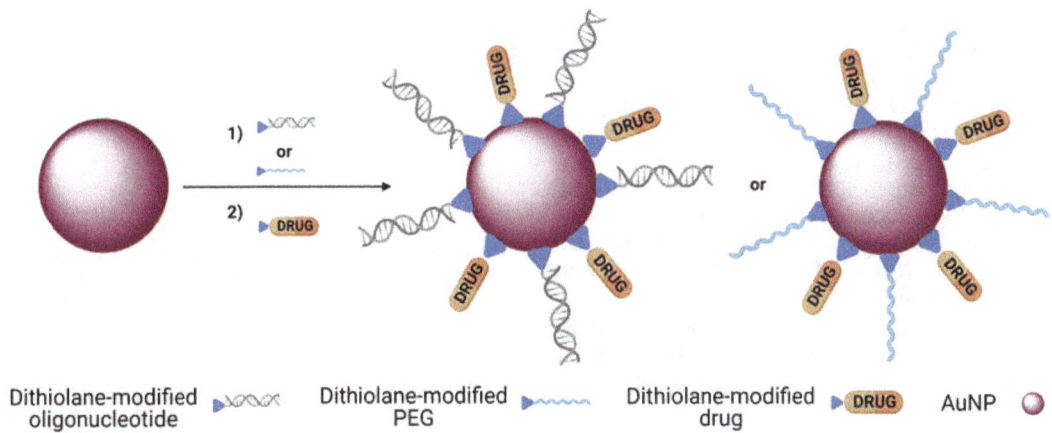

Fig. 4 Schematic representation of AuNPs functionalization with a dithiolane-modified drug using oligonucleotides or PEG as stabilizers

4. To get a complete functionalization of the nanoparticles, we recommend adding 10,000 pmol of the oligonucleotide to 1 mL AuNP (12 nM, 13 nm).

5. If you see that the AuNPs are being attached to the plastic tubes, move the solution to other plastic tubes immediately.

6. When AuNPs are changing their color to blue, it is due to aggregation. If vortexing the solution does not re-solubilize them, discard the preparation.

7. Oligonucleotide quantification could also be done by releasing the attached oligonucleotide [27]. To do so, treat the sample with 1 mM GSH for 8 h 37 °C. Then centrifuge the sample for

30 min at 13.2 rpm. Collect the supernatant and measure it as described in Subheading 3.3 and 3.4.

8. Drug quantification could also be done by comparing the drug supernatant absorbance or fluorescence with a proper standard linear calibration curve of the drug [19] or using analytical chromatography (HPLC).

References

1. Milán Rois P, Latorre A, Rodriguez Diaz C et al (2018) Reprogramming cells for synergistic combination therapy with nanotherapeutics against uveal melanoma. Biomimetics 3:28. https://doi.org/10.3390/biomimetics3040028

2. Hammond SM, Aartsma-Rus A, Alves S et al (2021) Delivery of oligonucleotide-based therapeutics: challenges and opportunities. EMBO Mol Med 13:e13243. https://doi.org/10.15252/emmm.202013243

3. Godfrey C, Desviat LR, Smedsrød B et al (2017) Delivery is key: lessons learnt from developing splice-switching antisense therapies. EMBO Mol Med 9:545–557. https://doi.org/10.15252/emmm.201607199

4. Bregoli L, Movia D, Gavigan-Imedio JD et al (2016) Nanomedicine applied to translational oncology: a future perspective on cancer treatment. Nanomed Nanotechnol Biol Med 12:81–103. https://doi.org/10.1016/j.nano.2015.08.006

5. Gou Y, Miao D, Zhou M et al (2018) Bio-inspired protein-based nanoformulations for cancer theranostics. Front Pharmacol 9:421–462. https://doi.org/10.3389/fphar.2018.00421

6. Mahajan S, Patharkar A, Kuche K et al (2018) Functionalized carbon nanotubes as emerging delivery system for the treatment of cancer. Int J Pharm 548:540–558. https://doi.org/10.1016/j.ijpharm.2018.07.027

7. Singh P, Pandit S, Mokkapati VRSS et al (2018) Gold nanoparticles in diagnostics and therapeutics for human cancer. Int J Mol Sci 19:1979. https://doi.org/10.3390/ijms19071979

8. Mishra DK, Shandilya R, Mishra PK (2018) Lipid based nanocarriers: a translational perspective. Nanomed Nanotechnol Biol Med 14:2023–2050. https://doi.org/10.1016/j.nano.2018.05.021

9. Núñez C, Estévez SV, del Pilar CM (2018) Inorganic nanoparticles in diagnosis and treatment of breast cancer. J Biol Inorg Chem 23:331–345. https://doi.org/10.1007/s00775-018-1542-z

10. Manzano M, Vallet-Regí M (2018) Mesoporous silica nanoparticles in nanomedicine applications. J Mater Sci Mater Med 29:65. https://doi.org/10.1007/s10856-018-6069-x

11. Pattni BS, Chupin VV, Torchilin VP (2015) New developments in liposomal drug delivery. Chem Rev 115:10938–10966. https://doi.org/10.1021/acs.chemrev.5b00046

12. Steinmetz NF (2013) Viral nanoparticles in drug delivery and imaging. Mol Pharm 10:1–2. https://doi.org/10.1021/mp300658j

13. Daraee H, Eatemadi A, Abbasi E et al (2016) Application of gold nanoparticles in biomedical and drug delivery. Artif Cells Nanomed Biotechnol 44:410–422. https://doi.org/10.3109/21691401.2014.955107

14. Turkevich J, Stevenson PC, Hillier J (1951) A study of the nucleation and growth processes in the synthesis of colloidal gold. Discuss Faraday Soc 11:55. https://doi.org/10.1039/df9511100055

15. Cutler JI, Auyeung E, Mirkin CA (2012) Spherical nucleic acids. J Am Chem Soc 134:1376–1391. https://doi.org/10.1021/ja209351u

16. Mokhtarzadeh A, Vahidnezhad H, Youssefian L et al (2019) Applications of spherical nucleic acid nanoparticles as delivery systems. Trends Mol Med 25(12):1066–1079. https://doi.org/10.1016/j.molmed.2019.08.012

17. Mioc A, Mioc M, Ghiulai R et al (2019) Gold nanoparticles as targeted delivery systems and theranostic agents in cancer therapy. Curr Med Chem 26(35):6493–6513. https://doi.org/10.2174/0929867326666190506123721

18. Heo DN, Yang DH, Moon H-J et al (2012) Gold nanoparticles surface-functionalized with paclitaxel drug and biotin receptor as theranostic agents for cancer therapy. Biomaterials 33:856–866. https://doi.org/10.1016/j.biomaterials.2011.09.064

19. Latorre A, Posch C, Garcimartín Y et al (2014) DNA and aptamer stabilized gold nanoparticles for targeted delivery of anticancer

therapeutics. Nanoscale 6:7436–7442. https://doi.org/10.1039/C4NR00019F

20. Kapadia CH, Melamed JR, Day ES (2018) Spherical nucleic acid nanoparticles: therapeutic potential. BioDrugs 32:297–309. https://doi.org/10.1007/s40259-018-0290-5

21. Randeria PS, Seeger MA, Wang X-Q et al (2015) siRNA-based spherical nucleic acids reverse impaired wound healing in diabetic mice by ganglioside GM3 synthase knockdown. Proc Natl Acad Sci 112:5573–5578. https://doi.org/10.1073/pnas.1505951112

22. Wang X, Hao L, Bu HF et al (2016) Spherical nucleic acid targeting microRNA-99b enhances intestinal MFG-E8 gene expression and restores enterocyte migration in lipopolysaccharide-induced septic mice. Sci Rep 6:1–13. https://doi.org/10.1038/srep31687

23. Latorre A, Posch C, Garcimartín Y et al (2014) Single-point mutation detection in RNA extracts using gold nanoparticles modified with hydrophobic molecular beacon-like structures. Chem Commun 50(23):3018–3020. https://doi.org/10.1039/c3cc47862a

24. Coutinho C, Somoza Á (2019) MicroRNA sensors based on gold nanoparticles. Anal Bioanal Chem 411(9):1807–1824

25. Posch C, Latorre A, Crosby MB et al (2015) Detection of GNAQ mutations and reduction of cell viability in uveal melanoma cells with functionalized gold nanoparticles. Biomed Microdev 17:15–37. https://doi.org/10.1007/s10544-014-9908-7

26. Liu X, Atwater M, Wang J, Huo Q (2007) Extinction coefficient of gold nanoparticles with different sizes and different capping ligands. Colloids Surf B Biointerfaces 58:3–7. https://doi.org/10.1016/J.COLSURFB.2006.08.005

27. Melamed JR, Riley RS, Valcourt DM et al (2017) Quantification of siRNA duplexes bound to gold nanoparticle surfaces. Methods Mol Biol 1570:1–15

3

Rapid Determination of MBNL1 Protein Levels by Quantitative Dot Blot for the Evaluation of Antisense Oligonucleotides in Myotonic Dystrophy Myoblasts

Nerea Moreno, Irene González-Martínez, Rubén Artero and Estefanía Cerro-Herreros

Abstract

Western blot assays are not adequate for high-throughput screening of protein expression because it is an expensive and time-consuming technique. Here we demonstrate that quantitative dot blots in plate format are a better option to determine the absolute contents of a given protein in less than 48 h. The method was optimized for the detection of the Muscleblind-like 1 protein in patient-derived myoblasts treated with a collection of more than 100 experimental oligonucleotides.

Key words Myotonic dystrophy, Oligonucleotides, Quantitative dot blot, Muscleblind-like protein 1, DM1 myoblasts

1 Introduction

Myotonic dystrophy type 1 (DM1) is a degenerative genetic disease that is classified as rare because it affects less than 1 in 2000 people (1/3000 to 1/8000; [1]). DM1 originates from an expansion of the CTG trinucleotide repeat in the $3'$-untranslated region (UTR) of the DMPK gene that, upon transcription, forms CUG hairpins that behave as toxic RNAs. CUG expansion RNA aberrantly binds and sequesters essential developmental proteins of the Muscleblind-like (MBNL) family, which are key regulators of alternative splicing. The depletion in MBNL protein function causes alterations in RNA metabolism that originate defined symptoms of the disease [2]. Studies in animal models have shown that the increase of MBNL in a genetic background of DM improves the

Nerea Moreno and Irene González-Martínez contributed equally to this work.

pathological phenotypes and that the overexpression of MBNL1 in control mice is well tolerated [3].

Patients suffer from myotonia and muscle atrophy and weakness, which, in advanced stages of the disease, lead to respiratory distress and early death. Currently, there is no effective treatment for DM1, and management of symptoms is the only option to preserve the quality of life of people living with DM1. In its most common form, the onset of symptoms occurs during adolescence, and affected people have a significantly shortened lifespan of 48–55 years. Because of its prevalence and the severity of the clinical manifestations, finding a cure for DM1 is a social and medical need [4, 5].

The development of effective high-throughput tools in drug discovery research has increased the demand for complementary high capacity immunoblot methods in which to assess the consequences of drug candidates at protein level. One example is the need to quickly evaluate the levels of MBNL1 protein in patient-derived myoblasts [6] treated with hundreds of oligonucleotide variants to block repressive miRNAs miR-23b and miR-218, as a means to boost endogenous levels and compensate sequestration by CUG expansions in mutant DMPK [7]. To this end, we have generated a diversity (>100) of highly modified antisense oligonucleotides (AONs) to block miR-23b and miR-218. Examples of these modifications are the substitution of natural ribose rings by locked nucleic acid (LNAs), C2′ hydroxyl substitutions with a methoxy (2′OMe) or methoxyethyl (2′-MOE), or the use of phosphorothioates to link two nucleotides, instead of natural phosphodiester bonds, to improve stability in vivo as they make these ASOs resistant to intracellular and extracellular nucleases. AON can also be made electrostatically neutral by using phosphorodiamidate morpholino oligomers (PMO) and can be conjugated to a cholesterol moiety to improve the diffusion through cell membranes and cell uptake [8, 9].

For this purpose, we propose the use of quantitative dot blot (QDB) analysis as an alternative to Western blot. For the development of the QDB assay, we have modified two previously published protocols [10, 11]. Dot blot was developed to simplify the process of Western blot analysis (Fig. 1) when the antibody is very specific for the detection of the protein of interest, and there is no need to determine its molecular weight, for example, when screening the effects of several molecules on the expression of a single protein. Specifically, QDB transforms traditional immunoblots into quantitative assays and allows expression analysis of a certain protein in your samples in 96-well format, being more efficient and faster than a Western blot.

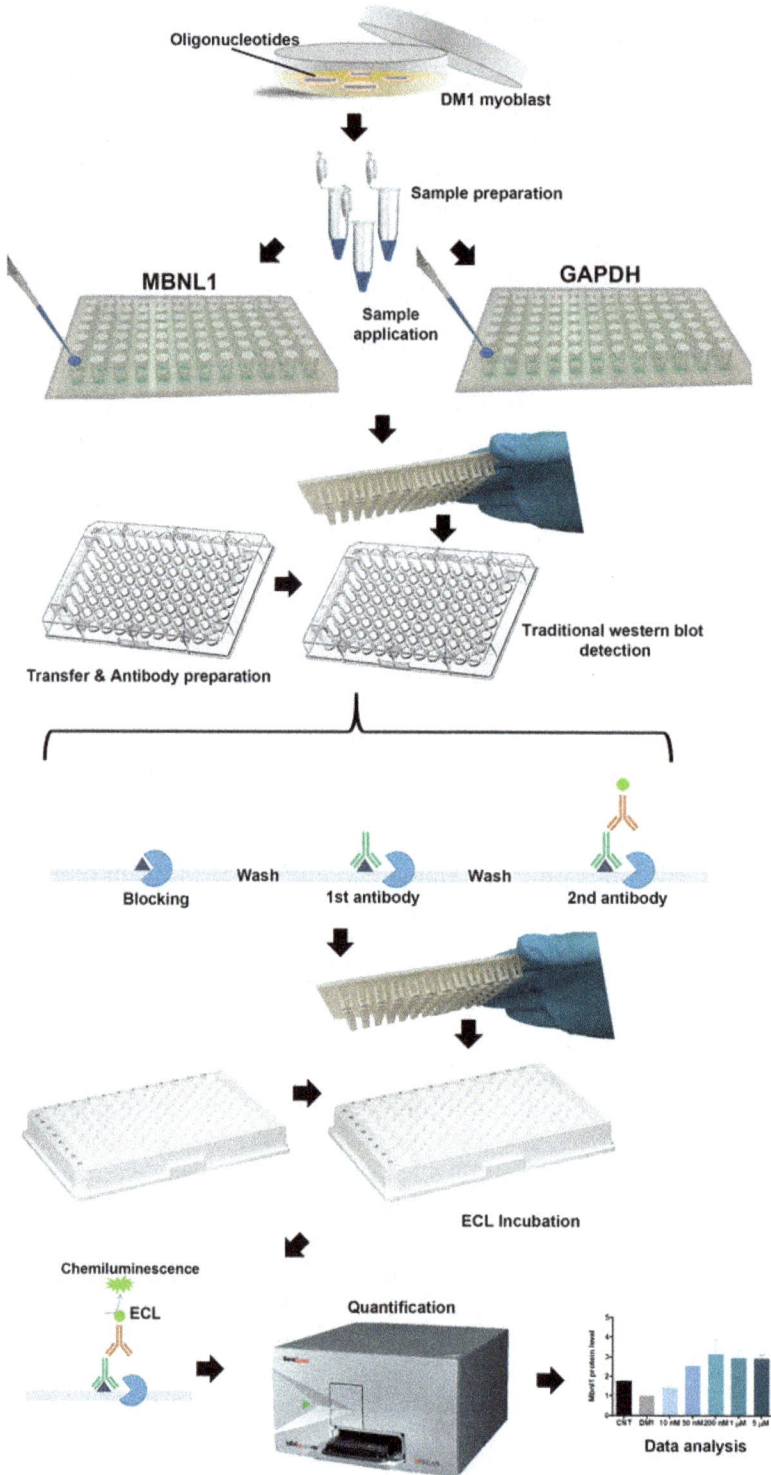

Fig. 1 Illustration of the entire QDB process for evaluation of antisense oligonucleotides in DM1 myoblast

2 Materials

2.1 Cell Culture and Transfection

1. Standard tissue culture facilities.
2. Six-well plates.
3. PBS 1×.
4. Opti-MEM Reduced Serum medium.
5. Transfection Reagent (*see* **Note 1**).
6. Complete medium (for 500 mL): 445 mL DMEM Dulbecco's Modified Eagle Medium 4500 mg/glucose, 50 mL FBS, and 5 mL penicillin-streptomycin.
7. Differentiation medium (for 100 mL): 1 mL penicillin-streptomycin, 2 mL Horse Serum, 1 mL apotransferrin, 100 μL insulin, 20 μL doxiciclin and 95.88 mL DMEM 4500 mg/glucose.

2.2 QDB Assay

1. Protein quantification kit (*see* **Note 2**).
2. QDB plates (Quanticision Diagnostics, Inc.).
3. Primary and secondary antibodies:
    (a) Anti-MBNL1 (Abcam, ab77017).
    (b) Anti-GAPDH (Santa Cruz Biotechnology, G-9: sc-365062).
    (c) Anti-mouse POD.
4. Pierce™ ECL Western blotting substrate.
5. White 96-well plate, flat bottom.
6. TECAN infinite M200 Pro.
7. Orbital shaker.
8. Transfer buffer: 14.4 g glycine, 3 g Tris base, 700 mL ddH$_2$O, 200 mL methanol and fill to 1 L with ddH$_2$O. Store at 4 °C.
9. 10× TBS: 24.2 g Tris base, 80 g NaCl, bring to 0.8 L with ddH$_2$O, adjust the pH to 7.6 with HCl and fill to 1 L with ddH$_2$O. Store at 4 °C.
10. 1× TBST: 100 mL 10× TBS, 5 mL Tween 20 and fill to 1 L with ddH$_2$O. Store at 4 °C.
11. Blocking buffer: 5 g milk powder and bring to 100 mL with TBST 1×. Store at 4 °C.
12. 4× Loading buffer: 0.8 mL 3 M Tris-HCl pH 6.8, 4 mL glycerol, 0.8 g SDS, 0.04 g bromophenol blue, 4 mL β-mercaptoethanol and fill to 10 mL with ddH$_2$O. Store at −20 °C.
13. RIPA solution: 10 mL RIPA and 1 tablet of protease inhibitor cocktail. Storage in 4 °C.

3 Methods

3.1 Cell Transfection

1. Seed cells in 6-well plates at a density of 200,000 cells per well in complete medium and are incubated for 24 h at 37 °C 5% CO_2.

2. Prepare the mixes for transfection. Volumes in this example are calculated for 3 wells of a 6-well plate and in 5 different concentrations in the range between 10 and 5000 nM, diluted in Opti-MEM in a sterile tube in a final volume of 600 μL. For example, to transfect AO at 200 μM mix 594 μL of Opti-MEM and add 6 μL of AO at 100 μM. To test other concentrations, adapt the volumes accordingly. Pipet gently to mix.

3. Add 3 μL of X-tremeGENE HP DNA transfection reagent to the diluted mix. Incubate for 30 min at RT while preparing the following step.

4. Remove medium from wells and wash twice with PBS 1×.

5. Add 200 μL of transfection mix to each well in a dropwise manner. Swirl the wells to ensure the distribution over the entire plate.

6. Leave the cells in the incubator for 4 h.

7. Add 2.3 mL of differentiation medium per well.

3.2 Sample Collection and Quantification

1. After 72 h of incubation of the DM1 myoblasts with the test oligonucleotides, wash each well with PBS before protein collection.

2. Collect cells in RIPA solution (120 μL/well).

3. Lyse cells with an ultrasonic homogenizer such as the UP100H. Sonication is at an ultrasonic cycle mode of 30 s and an amplitude of 60%. The sample must be kept in ice during the entire procedure.

4. Centrifuge the lysate at $16,000 \times g$ for approximately 15 min after ultrasonic homogenization/extraction.

5. After centrifuging, collect the supernatant and determine the protein concentration by a protein assay such as Pierce protein assay BCA.

3.3 Sample Preparation

1. Calculate enough protein extract for ten replicate wells at 1 μg/well to account for pipetting errors, as they will be later loaded in quadruplicates in two different plates: one for detection of MBNL1 and the other for GAPDH, as an endogenous control (see **Note 3**).

2. Add the protein extract at the indicated concentration, add 10.4 μL of loading buffer 4× and complete to a final volume of 50 μL of ddH_2O (see **Note 4**).

3. Boil the freshly prepared sample for 5 min in a water bath and after denaturation of the protein leave it on ice.

3.4 Sample Application and Transference

1. Place the plates upside down to load the samples. Pipette 5 μL of the previously prepared protein sample on each membrane circle (Fig. 2).

2. Dry loaded QDB plates for 30 min at room temperature in a well-ventilated space.

3. After drying, dip the QDB plate in transfer buffer for 1 min. During this minute, shake it with an orbital movement until the blue color in the wells is eliminated.

4. Rinse QDB plates gently with TBST in constant shaking for 1 min three times.

5. Block QDB plates with blocking buffer for 1 h with constant shaking.

3.5 Primary Antibody Incubation

1. Dilute primary antibodies in blocking buffer at the appropriate concentrations (anti-MBNL1: 1/1000 and anti-GAPDH: 1/500) and aliquot into the wells of two 96-well plates, one for each antibody, at 100 μL/well (*see* **Note 5**).

2. Insert each QDB plate into the 96-well plates and incubate overnight at 4 °C in constant shaking.

3. Rinse the plates briefly with TBST and then wash with TBST in constant shaking for 5 min, three times.

3.6 Secondary Antibody Incubation

1. Dilute the appropriate secondary antibody (anti-mouse POD) in blocking buffer at the final concentration of 1/2000 and aliquot it in the 96-well plates at 100 μL/well. QDB plates are inserted into the 96-well plates and incubated in constant shaking for 2 h.

2. Rinse QDB plates gently three times with TBST and then wash with TBST in constant shaking for 5 min, three times.

3.7 Quantification

1. Before the last washes, prepare the ECL substrate by following the manufacturer's instructions.

2. Aliquot 50 μL of ECL substrate into each well of a 96-well plate, and insert the QDB plate inside the 96-well plates for 1 min.

3. After a minute, remove the QDB plate from the 96-well plate and shake it briefly to remove the remaining reagent and introduce it into a clean white 96-well plate.

4. For data acquisition, use a TECAN infinite M200 Pro using an initial acquisition time of 1 s/well in the luminescence and select "plate with cover."

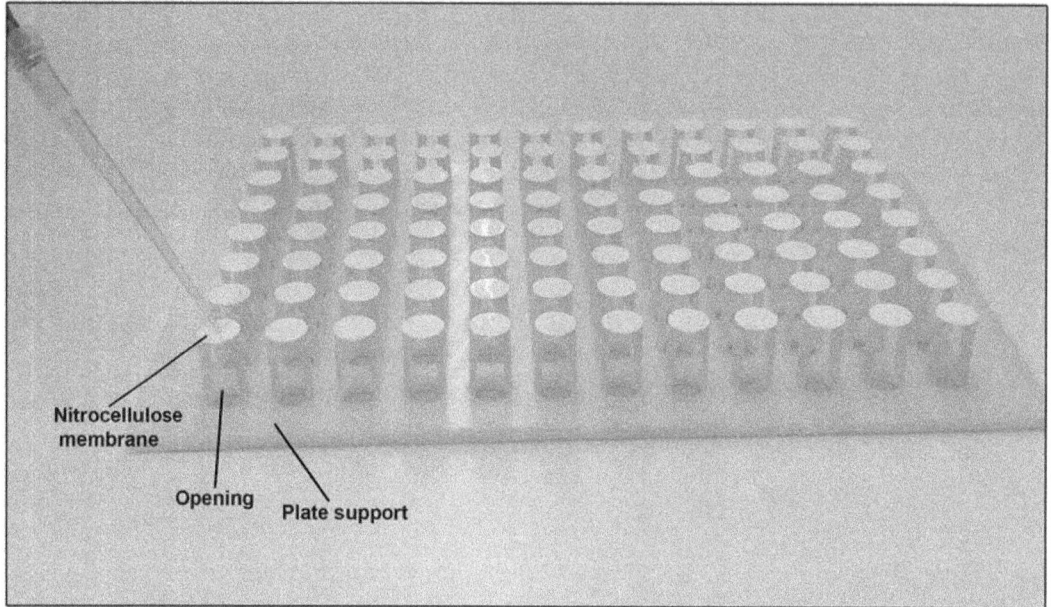

Fig. 2 A photograph of a QDB plate upside down. The figure shows the different components of the plate and how to load the sample

3.8 Data Analysis

1. First, look at the negative control values, which have to be below the rest of the samples. Pay also attention to the integrity of the plate since any alteration of the membrane in one of the wells may lead to artifactually low chemiluminescence values.

2. Next, subtract the average value of the negative controls from all the samples, thus removing the contribution of the background to the quantification.

3. Obtain the average of the quadruplicates for both the MBNL1 and the GAPDH plates. Divide the average of the replicates of the MBNL1 samples by those of GAPDH.

4. GAPDH normalized values can be further normalized to values in mock-transfected DM1 cells to obtain fold changes.

5. As a positive control, confirm in each plate that the difference between the average values of MBNL1 in control and untreated DM1 cells is around 1.6 times.

4 Notes

1. Could be done with any transfection reagent, but we have tested only X-tremeGENE HP DNA Transfection Reagent.

2. Could be done with any protein quantification kit, but we have tested only PierceTM BCA Protein Assay kit.

3. Always include positive and negative controls in each of the plates. Our positive controls were healthy cell samples while the negative contained RIPA buffer only. Negative control wells provide the background luminescence reading, which must be significantly lower than our experimental data.

4. It is very important to prepare the protein and load the samples in the gas hood since β-mercaptoethanol is toxic by inhalation.

5. To ensure a valid result, the specificity of the antibody used needs to be verified through Western blot analysis.

Acknowledgments

The project leading to these results has received funding from "la Caixa" Banking Foundation under the project code HR17-00268 (TATAMI project to R.A.). I.G.-M. was funded by the Precipita Project titled "Desarrollo de una terapia innovadora contra la distrofia miotónica"; E.C.-H. was supported by the post-doctoral fellowship APOSTD/2019/142; and N.M.-C was supported by the pre-doctoral fellowship PRE2019-090622. Part of the equipment employed in this work has been funded by Generalitat Valenciana and cofinanced with ERDF funds (OP ERDF of Comunitat Valenciana 2014-2020).

References

1. Ashizawa T, Gagnon C, Groh WJ, Gutmann L, Johnson NE, Meola G, Moxley R III, Pandya S, Rogers MT, Simpson E, Angeard N, Bassez G, Berggren KN, Bhakta D, Bozzali M, Broderick A, Byrne JLB, Campbell C, Cup E, Day JW, De Mattia E, Duboc D, Duong T, Eichinger K, Ekstrom AB, van Engelen B, Esparis B, Eymard B, Ferschl M, Gadalla SM, Gallais B, Goodglick T, Heatwole C, Hilbert J, Holland V, Kierkegaard M, Koopman WJ, Lane K, Maas D, Mankodi A, Mathews KD, Monckton DG, Moser D, Nazarian S, Nguyen L, Nopoulos P, Petty R, Phetteplace J, Puymirat J, Raman S, Richer L, Roma E, Sampson J, Sansone V, Schoser B, Sterling L, Statland J, Subramony SH, Tian C, Trujillo C, Tomaselli G, Turner C, Venance S, Verma A, White M, Winblad S (2018) Consensus-based care recommendations for adults with myotonic dystrophy type 1. Neurol Clin Pract 8(6):507–520. https://doi.org/10.1212/CPJ.0000000000000531

2. Jiang H, Mankodi A, Swanson MS, Moxley RT, Thornton CA (2004) Myotonic dystrophy type 1 is associated with nuclear foci of mutant RNA, sequestration of muscleblind proteins and deregulated alternative splicing in neurons. Hum Mol Genet 13(24):3079–3088. https://doi.org/10.1093/hmg/ddh327

3. Chamberlain CM, Ranum LP (2012) Mouse model of muscleblind-like 1 overexpression: skeletal muscle effects and therapeutic promise. Hum Mol Genet 21(21):4645–4654. https://doi.org/10.1093/hmg/dds306

4. Landfeldt E, Nikolenko N, Jimenez-Moreno C, Cumming S, Monckton DG, Gorman G, Turner C, Lochmuller H (2019) Disease burden of myotonic dystrophy type 1. J Neurol 266(4):998–1006. https://doi.org/10.1007/s00415-019-09228-w

5. Mathieu J, Allard P, Potvin L, Prevost C, Begin P (1999) A 10-year study of mortality in a cohort of patients with myotonic dystrophy. Neurology 52(8):1658–1662

6. Arandel L, Polay Espinoza M, Matloka M, Bazinet A, De Dea DD, Naouar N, Rau F, Jollet A, Edom-Vovard F, Mamchaoui K, Tarnopolsky M, Puymirat J, Battail C, Boland A, Deleuze JF, Mouly V, Klein AF, Furling D (2017) Immortalized human

myotonic dystrophy muscle cell lines to assess therapeutic compounds. Dis Model Mech 10(4):487–497. https://doi.org/10.1242/dmm.027367

7. Cerro-Herreros E, Sabater-Arcis M, Fernandez-Costa JM, Moreno N, Perez-Alonso M, Llamusi B, Artero R (2018) miR-23b and miR-218 silencing increase Muscleblind-like expression and alleviate myotonic dystrophy phenotypes in mammalian models. Nat Commun 9(1):2482. https://doi.org/10.1038/s41467-018-04892-4

8. Crooke ST, Witztum JL, Bennett CF, Baker BF (2019) RNA-targeted therapeutics. Cell Metab 29(2):501. https://doi.org/10.1016/j.cmet.2019.01.001

9. Smith CIE, Zain R (2019) Therapeutic oligonucleotides: state of the art. Annu Rev Pharmacol Toxicol 59:605–630. https://doi.org/10.1146/annurev-pharmtox-010818-021050

10. Tian G, Tang F, Yang C, Zhang W, Bergquist J, Wang B, Mi J, Zhang J (2017) Quantitative dot blot analysis (QDB), a versatile high throughput immunoblot method. Oncotarget 8(35):58553–58562. https://doi.org/10.18632/oncotarget.17236

11. Qi X, Zhang Y, Zhang Y, Ni T, Zhang W, Yang C, Mi J, Zhang J, Tian G (2018) High throughput, absolute determination of the content of a selected protein at tissue levels using quantitative dot blot analysis (QDB). J Vis Exp (138):56885. https://doi.org/10.3791/56885

4

Modeling Splicing Variants Amenable to Antisense Therapy by use of CRISPR-Cas9-Based Gene Editing in HepG2 Cells

Arístides López-Márquez, Ainhoa Martínez-Pizarro, Belén Pérez, Eva Richard and Lourdes R. Desviat ⓘ

Abstract

The field of splice modulating RNA therapy has gained new momentum with FDA approved antisense-based drugs for several rare diseases. In vitro splicing assays with minigenes or patient-derived cells are commonly employed for initial preclinical testing of antisense oligonucleotides aiming to modulate splicing. However, minigenes do not include the full genomic context of the exons under study and patients' samples are not always available, especially if the gene is expressed solely in certain tissues (e.g. liver or brain). This is the case for specific inherited metabolic diseases such as phenylketonuria (PKU) caused by mutations in the liver-expressed *PAH* gene.

Herein we describe the generation of mutation-specific hepatic cellular models of PKU using CRISPR/Cas9 system, which is a versatile and easy-to-use gene editing tool. We describe in detail the selection of the appropriate cell line, guidelines for design of RNA guides and donor templates, transfection procedures and growth and selection of single-cell colonies with the desired variant, which should result in the accurate recapitulation of the splicing defect.

Key words Splicing, Gene editing, CRISPR/Cas9, HepG2, Inherited metabolic diseases, Phenylketonuria, Cellular models

1 Introduction

Splicing defects account for up to one-third of human disease-causing variants, according to the current estimates [1–3]. Constitutive splicing relies on the recognition of consensus splicing sequences (5′ splice site, 3′ splice site, branch point, and polypyrimidine tract) by spliceosomal components, as well as of other less conserved regulatory elements, referred to as exonic or intronic splicing enhancers or silencers (ESE, ISE, ESS, or ISS), that modulate spliceosome recruitment [4]. These cis-regulatory elements are recognized by trans-acting factors including the serine/arginine-rich domain-containing (SR) protein and heterogeneous nuclear ribonucleoprotein (hnRNP) families that may act co-ordinately to accurately regulate exon inclusion.

Pathogenic splicing variants disrupt conserved splice sites or regulatory elements or cause aberrant splicing by creating/activating alternative splice sites or by promoting the aberrant inclusion of intronic pseudoexons [4]. Splicing can be modulated therapeutically using antisense approaches, and to date, the clinically approved splice-switching antisense oligonucleotides (SSO) for spinal muscular atrophy, Duchenne muscular atrophy and for an individual patient with a rare, fatal neurodegenerative disease [5–7], represent landmarks in the field, opening new avenues for treatment of patients with defects amenable to splice-mediated correction.

The first requirement for the accurate design and testing of antisense splice correction therapy is the availability of relevant experimental models in which to dissect the underlying molecular mechanisms of pathogenic variants and to test candidate molecules. In this sense, the development of clustered-regulatory interspaced short palindromic repeats (CRISPR)-CRISPR associated nuclease (Cas) genome editing has paved the way to the rapid and easy generation of new and improved cell/animal models of disease. This has facilitated the understanding of the specific pathogenic effect and has allowed efficient testing of targeted therapies, including allele-specific repair for splicing mutations, in tissue types with native expression levels [8–13]. Based on a naturally employed bacterial defense mechanism [14, 15], CRISPR/Cas9 technology was developed as a programmable system of genetic editing that commonly uses the Cas9 nuclease from *Streptococcus pyogenes* and a RNA duplex comprised of a sequence-specific CRISPR RNA (crRNA) and a generic trans-activating CRISPR RNA (tracrRNA) that directs the nuclease to a cut site point, three base pairs upstream of the protospacer adjacent motif or PAM. The PAM is a three-nucleotide motif essential for the nuclease to recognize its DNA target which in the case of Cas9 is NGG. The crRNA and the tracrRNA can be delivered individually or linked in a single RNA molecule. These elements can be delivered to cells as plasmids or as a ribonucleoprotein (RNP) complex [16].

Once Cas9 nuclease cuts the DNA introducing a double stranded break (DSB), the cell can repair this through two different mechanisms: non-homologous end-joining (NHEJ) which usually results in small insertions or deletions, useful for the generation of gene knockouts, or homology driven repair (HDR), used to introduce specific changes via a DNA template with homology arms to our target locus and containing the sequence or point mutation desired [16].

In our laboratory we have used CRISPR/Cas9 technology to introduce splicing mutations causing inherited metabolic diseases (IMD) in cellular and animal models. IMD are monogenic diseases characterized by dysregulation of the metabolic networks that underlie development and homeostasis [17]. They belong to the category of rare diseases due to their low individual prevalence and are generally enzyme deficiencies of autosomal recessive inheritance, characterized by the toxic accumulation of precursors and of their derivatives or by lack of downstream metabolites. Several of the most frequent and well characterized IMD, e.g. organic acidemias and amino acid disorders, are of major hepatic expression and, as in other genetic diseases, 13–25% of all disease-causing variants interfere with mRNA splicing (HGMD statistics, Professional Release 2019.3). These data warrant further investigation of the therapeutical potential of SSOs in these diseases and the generation of liver specific cellular models for these studies.

The generation of a cell model using CRISPR/Cas9 system can be done in a huge variety of cell lines. In this chapter we describe the protocol for efficient introduction of a specific splicing variant in the *PAH* gene, coding for phenylalanine hydroxylase, and responsible for the well characterized disease phenylketonuria (PKU, MIM#261600), inherited in autosomal recessive fashion. Human *PAH* is exclusively expressed in liver, so in this protocol we use hepatoma cell line HepG2 seeking to attain edition in both alleles (homozygous phenotype). We explain how to select for the appropriate cell line in each particular case, describe the design of RNA guides and donor templates, transfection procedures, growth of single-cell colonies, selection and testing to confirm genomic edition, and accurate recapitulation of the splicing defect (Fig. 1). Appropriate controls to be included in each step are explained, as well as the necessary precautions to be taken especially for intronic splicing variants. We use as example the CRISPR/Cas9-mediated introduction of the recently characterized *PAH* intronic variant, c.1199 + 20G > C, that causes exon skipping due to disruption of a splicing regulatory element [18]. This variant creates a PshAI restriction site, which is used to screen for gene edition in the transfected cells.

2 Materials

2.1 Cell Culture

1. Laminar flow-hood.
2. Humid CO_2 incubator.
3. Centrifuge.
4. Phase-contrast microscope.
5. Hemocytometer–double chamber with Neubauer rulings.
6. Manual Counter.
7. Consumables: Tissue culture plates, filtered tips, falcon tubes, Eppendorf tubes.

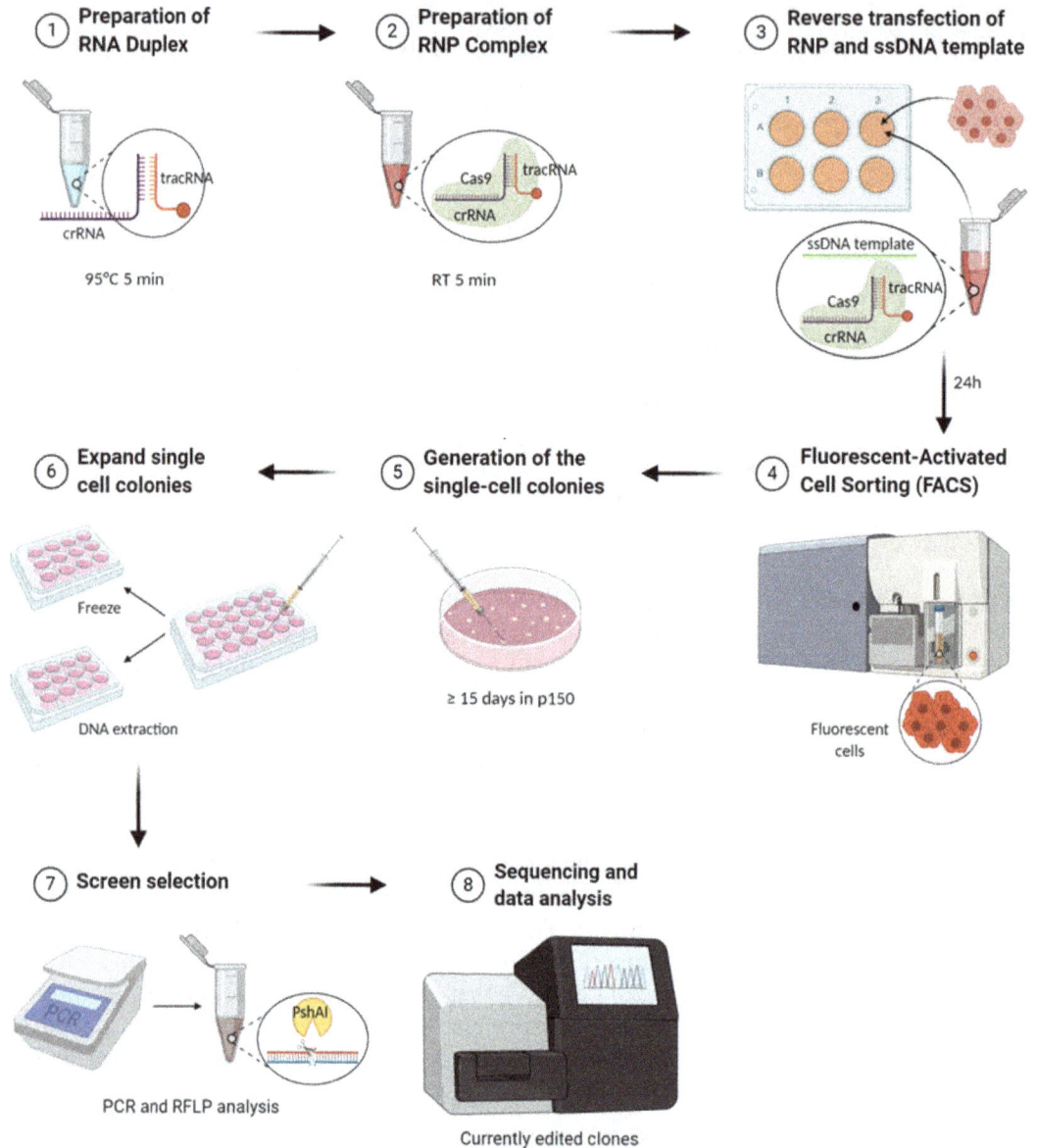

Fig. 1 Outline of the gene editing experimental protocol. (This image was created using BioRender)

8. Phosphate Buffered Saline (PBS).

9. Minimum Essential Medium (MEM) supplemented with 10% Fetal Bovine Serum (FBS), 1% L-Glutamine, and antibiotics.

10. Solution of trypsin–EDTA: 0.25% trypsin, 1 mM EDTA.

11. Trypan Blue Solution: 0.4% trypan blue in PBS.

12. Micropipettes.

13. Stripper micropipettes and 150 μm tips (Origio Inc).

2.2 Ribonucle-oprotein (RNP) Transfection	1. Cas9 Nuclease (*see* **Note 1**).
	2. Fluorescently labeled tracrRNA (tracrRNA-ATTO550) (*see* **Note 1**).
	3. Single-stranded (ss) DNA Template (*see* **Note 1**).
	4. crRNA (*see* **Note 1**).
	5. RNA Duplex Buffer supplied by the manufacturer (*see* **Note 1**).
	6. Nuclease-Free Water.
	7. OptiMEM media.
	8. Lipofectamine Transfection Reagent (*see* **Note 2**).
2.3 Fluorescence Activated Cell Sorting	1. Sorting buffer: PBS, 5 mM EDTA, 25 mM Hepes pH 7.0 supplemented with 2% FBS.
	2. 5-mL polystyrene tubes with cell strainer.
	3. Cell Sorter.
2.4 Genomic DNA Isolation	1. QIAamp DNA Mini Kit for DNA purification (Qiagen).
	2. Centrifuge.
	3. NanoDrop One spectrophotometer (Thermo Fisher Scientific).
2.5 Polymerase Chain Reaction (PCR)	1. Thermal cycler.
	2. PCR tubes.
	3. Nuclease-Free Water.
	4. dNTPs.
	5. FastStart Taq DNA Polymerase (Roche) and PCR buffer $10\times$ (25 mM $MgCl_2$).
	6. Target-specific primers.
	7. Agarose gel with ethidium bromide (0.4 μg/mL) and UV transilluminator.
	8. DNA Molecular Weight Marker.
	9. Kit to purify PCR products, e.g., Cycle Pure Kit for PCR product purification (Omega).
2.6 Restriction Fragment Length Polymorphism Assay (RFLP)	1. PshAI restriction enzyme and enzyme reaction buffer.
	2. Agarose gel with ethidium bromide (0.4 μg/mL) and UV transilluminator.
	3. DNA Molecular Weight Marker.
2.7 RNA Isolation and Reverse Transcription	1. Trizol isolation reagent (Ambion).
	2. 2-Propanol and chloroform.
	3. Ethanol 75%.

4. RNase-Free Water and RNase free-consumables.

5. NanoDrop One spectrophotometer (Thermo Fisher Scientific).

6. Thermal cycler.

7. NZY First-Strand cDNA Synthesis Kit.

2.8 Web Resources

1. Sequences and genomes: https://www.ensembl.org/.

2. Sequences alignments: https://blast.ncbi.nlm.nih.gov/Blast.cgi.

3. Design and analysis of crRNAs: https://bioinfogp.cnb.csic.es/tools/breakingcas/.

4. Design, analysis and/or ordering of crRNAs, tracrRNA, ssDNA Templates: https://eu.idtdna.com/site/order/designtool/index/CRISPR_SEQUENCE.

5. PCR primer design: http://bioinfo.ut.ee/primer3-0.4.0/.

6. Primer and PCR product analysis: https://www.ncbi.nlm.nih.gov/tools/primer-blast/.

3 Methods

3.1 HepG2 Cell Culture

1. Culture the selected HepG2 cell line following standard procedures in P-100 culture dishes with MEM supplemented with 10% FBS, antibiotics, and glutamine at 37 °C in an incubator with 95% humidity and 5% CO_2 (*see* **Notes 3–6**).

2. Just before transfection (see Subheading 3.5 below) detach cells by trypsinization. First, aspirate the media and wash the cells with PBS. Once the PBS has been aspirated from the plate, add 0.25% trypsin-EDTA into the plate and incubate at 37 °C for 5 min. Check by microscopy that the cells are rounding up. Once the cells are detached from the plate add 10% FBS in MEM to stop the trypsin reaction. Pipette the cells up and down to dissociate detached cell clumps into single cells. Transfer the cells to a falcon tube and spin them in a centrifuge at 218 × *g* for 5 min. Discard the supernatant and resuspend the cell pellet in fresh medium.

3. Count the resuspended cells using a hemocytometer. Prepare a dilution 1:8 of the cells in Trypan Blue solution to distinguish dead cells (stained blue). Add the cell suspension to both chamber sides of the hemocytometer and count the cells with the help of a manual counter.

3.2 Design of Guide RNAs and Donor Template

1. Design the specific crRNA guides with the help of bioinformatics software and their potential off-targets (*see* **Notes 7–12**) (Fig. 2).

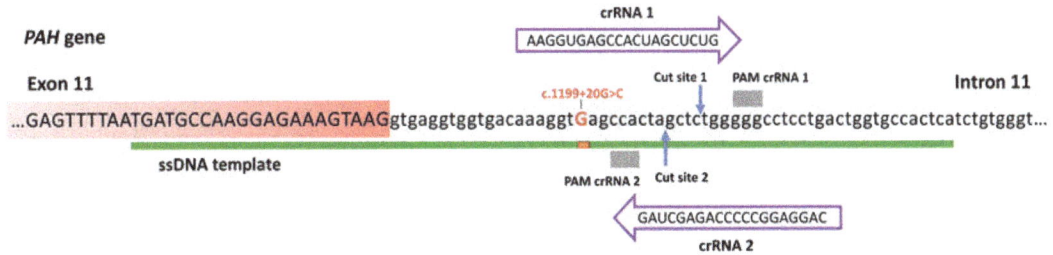

Fig. 2 Schematic representation of the *PAH* gene region surrounding the c.1199 + 20G > C mutation, showing the sequence of the ssDNA template which will include the desired change (green line with red box) and the crRNA guides used (purple arrows), indicating the corresponding PAM sequences (gray rectangles) and the Cas9 nuclease cut sites 3 nucleotides upstream of PAM (blue arrows)

2. Design the ssDNA donor template (*see* **Notes 13–15**) (Fig. 2).

3. Order the crRNA guides, tracrRNA, ssDNA donor template, Cas9 nuclease, transfection reagent, and all the necessary reagents for the CRISPR/Cas9 system (*see* **Note 1**).

3.3 Preparation of RNA Duplex

1. Resuspend the crRNA and tracrRNA-ATTO550 in 20 and 50 μL of Nuclease-Free Duplex Buffer, respectively, resulting in 100 μM stock concentrations (*see* **Notes 16–19**).

2. Prepare the RNA Duplex at a final concentration of 1 μM by mixing the tracrRNA and the crRNA in equimolar concentrations in Nuclease-Free Duplex Buffer (add 1 μL of each crRNA and tracRNA-ATTO550 to 98 μL of buffer) (*see* **Note 20**).

3. Heat at 95 °C for 5 min.

4. Cool to room temperature (25 °C).

3.4 Preparation of the Ribonucleoprotein Complex (RNP)

1. Dilute Cas9 nuclease to a working concentration of 1 μM in OptiMEM (*see* **Note 21**).

2. Prepare the RNP by mixing in independent tubes for each crRNA the following: 24 μL of RNA duplex (1 μM), 24 μL of Cas9 (1 μM), 9.6 μL of Cas9 PLUS reagent from CRISPR MAX kit (*see* **Note 21**), and 342.4 μL of OptiMEM, adding to a total 400 μL.

3. Incubate at room temperature for 5 min.

3.5 Reverse Transfection of RNP and DNA Donor Template

1. Prepare the ssDNA Donor Template at a working concentration of 1 μM in Nuclease-Free Water.

2. Prepare the transfection mixing the following for each well of a 6-well plate: 7.2 μL of 1 μM ssDNA donor template, 400 μL of RNP complex, 19.2 μL of CRISPRMAX Transfection reagent and 373.6 μL of OptiMEM, adding to a total 800 μL (*see* **Notes 22** and **23**).

3. Incubate at room temperature for 20 min.

4. During the incubation of the transfection mix proceed to detach HepG2 cells by trypsinization as explained above (Subheading 3.1) (*see* **Note 24**).

5. Prepare a dilution of 4×10^5 cells/mL with complete MEM without antibiotics.

6. Once the incubation of the transfection mix is complete, add 800 μL to each well of the 6-well plate.

7. Add 1600 μL of the cell suspension to each well containing the transfection mix, for a final volume of 2400 μL. The number of cells should be 6.4×10^5 cells/well; final concentration of RNP is 10 nM and final concentration of the ssDNA donor template is 3 nM (*see* **Note 22**).

8. Incubate the cells in an incubator at 37 °C and 5% CO_2 for 24 h. If you are not using tracrRNA-ATTO550 and performing FACS analysis, incubate for 48 h and skip (Subheading 3.6).

3.6 Fluorescent-Activated Cell Sorting (FACS)

1. Trypsinize the cells as explained above 24 h after transfection.

2. Dilute 1.5×10^6 cells in 300 μL of sorting buffer.

3. For each sample to be collected, a 15 mL Falcon tube with 2 mL of FBS supplemented with 2 μL of antibiotics mix must be prepared.

4. Collect fluorescent cells for each crRNA.

5. Centrifuge at $218 \times g$ for 5 min.

6. Seed 1×10^5 cells per well of 6-well plate. One complete 6-well plate for each crRNA is enough.

7. Incubate the cells in an incubator at 37 °C and 5% CO_2. Change the medium every 2 days. Expand the cell culture and freeze several cryotubes of the total pool of transfected cells (*see* **Note 25**).

3.7 Generation of the Single-Cell Colonies

1. Trypsinize the cells as explained above.

2. Count the cells with the help of a hemocytometer.

3. Seed 150–200 cells in a 150-mm plate.

4. Incubate the cells in an incubator at 37 °C and 5% CO_2 for, at least 15 days (*see* **Note 5**).

5. Once the colonies can be seen with the naked eye, select and pick the colonies with a stripper micropipette and 150 μm tips. This should be done by observing colonies under a microscope inside a laminar flow-hood under sterile conditions (*see* **Notes 26** and **27**).

6. Put each colony in a 1.5 mL Eppendorf tube with 50 μL of trypsin and incubate at 37 °C for 5 min.

7. Individualize the cells by pipetting up and down several times. Seed the cells derived from the colonies into 24-well plates. It is not necessary to centrifuge previously.

8. Expand the culture and change the medium every 48 h. Once the cells are confluent, trypsinize the cells and split them into two wells of a 12-well plate. One of the wells will be used to isolate DNA for analysis, while the other will be used to freeze and/or expand the colony (*see* **Notes 28** and **29**).

3.8 Genomic DNA Extraction and RFLP Analysis

1. Trypsinize the cells as indicated above, centrifuge the cells at $300 \times g$ for 5 min and discard the supernatant.

2. Resuspend the pellet with 200 μL of PBS.

3. Isolate the DNA using QIAamp DNA Mini kit (Qiagen) following the manufacturer's protocol (*see* **Note 30**).

4. Quantify the DNA concentration in the isolate using Nanodrop One spectrophotometer.

5. Design primers using Primer3 software (http://bioinfo.ut.ee/primer3-0.4.0/) for amplification of the region surrounding the desired edited change (500–600 bp) (*see* **Note 31**).

6. Prepare pools of DNA mixing equal amounts (circa 50 ng) of DNA from individual colonies (4 or 5) in a PCR tube to obtain a final amount of 200 ng (*see* **Note 32**).

7. Perform PCR according to standard procedures. Use 200 ng of genomic DNA in a 50 μL PCR reaction with 1 μM of each primer, 200 μM of each dNTP, 2 unit of Taq polymerase and PCR buffer 1×. The PCR amplification program is as follows: 1 cycle with 5 min at 95 °C, 36 subsequent cycles of 25 s at 95 °C, 25 s at 50–60 °C (depending on the primers), and 40 s at 72 °C, with a final 7-min extension at 72 °C.

8. Run 5 μL of each sample in a 2% agarose gel with ethidium bromide (or other safer dye, such as GelRed or SYBR Safe) to confirm amplification.

9. Digest 5 μL of each amplified sample in a final volume of 20 μL with the restriction enzyme using the appropriate buffer and following the manufacturer's indications (*see* **Notes 33** and **34**).

10. Run the restriction reaction volume in a 2% agarose gel with ethidium bromide to visualize the resulting DNA bands.

11. Repeat the PCR and the restriction assay for each individual clone included in the pools for which a positive RFLP analysis is observed (in our example, digestion with PhsAI) (Fig. 3).

A)   WILD-TYPE   MUTANT (c.1199+20G>C)

        tggtgacaaaggt**G**agcc   tggtgacaaaggt**C**agcc
            -PshAI   +PshAI

B)

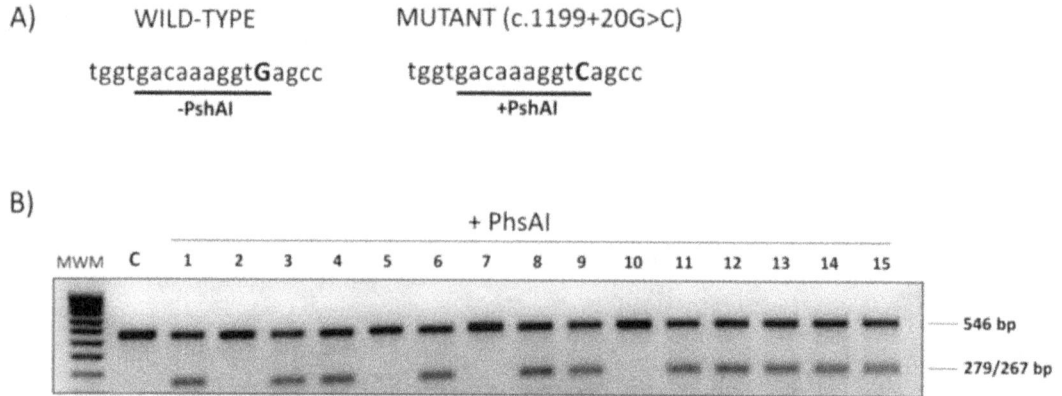

Fig. 3 RFLP analysis to monitor for gene edition. The wild-type and mutant sequences are shown in panel **a** and panel **b** is a representative gel showing RFLP analysis of single-cell colonies. Top bands correspond to the amplified PCR products and lower-sized bands correspond to the products obtained by digestion with PshAI enzyme due to the introduction of a restriction site with the point mutation c.1199 + 20G > C. *C* undigested control. Colonies 1, 3, 4, 6, 8, 9, 11–15 are positive and heterozygous (one allele edited)

3.9 Sequencing Analysis of Candidate Clones and Off-Targets Analysis

1. Using DNA from positive clones, perform a PCR to amplify the edited region and those regions where potential off-targets were identified by the software (*see* **Note 6**), using specific primers designed using Primer3 software (*see* **Note 35**).

2. Purify the PCR products using a PCR purification kit (*see* **Note 36**).

3. Prepare the mix for the sequencing reaction according to the instructions of the genomics facility and/or the sequencer. Carry out the sequencing with the forward and the reverse primers in separate reactions.

4. Analyze the sequences with the help of a chromatogram viewer (*see* **Notes 35** and **37**).

3.10 RNA Isolation

1. Once a correctly edited clone has been identified, expand the culture to obtain enough cells for RNA isolation.

2. Wash with PBS and trypsinize the cells. Centrifuge the cells at $16{,}000 \times g$ for 5 min and discard the supernatant. Cells can be frozen at $-70\,°C$ in this step.

3. Add 1 mL of Trizol per sample. Incubate the homogenate for 5 min at room temperature to achieve complete dissociation of nucleoprotein complexes.

4. Add 200 μL of chloroform. Mix by vortexing for 15 s and incubate for 2 min at room temperature.

5. Centrifuge at $12{,}000 \times g$ for 15 min at $4\,°C$ to separate the phases.

6. Transfer the aqueous phase (upper and transparent) to a new Eppendorf tube.

7. Add 500 μL of 2-propanol. Mix by vortexing for 15 s.

8. Incubate the samples at room temperature for 10 min followed by an incubation of, at least, 20 min at −20 °C.

9. Centrifuge the samples at 12,000 × g for 30 min at 4 °C. Discard the supernatant.

10. Add 1 mL of 75% ethanol and wash the precipitate by vortexing (it can be stored in 75% ethanol for a week at 4 °C or a year at −20 °C).

11. Centrifuge at 12,000 × g for 5 min at 4 °C and discard the supernatant.

12. Centrifuge at 7500 × g for 1 min at 4 °C. Let the pellet dry at room temperature until they become transparent.

13. Dissolve the dry RNA in 30 μL of "Nuclease-Free Water" by pipetting and incubate it at least 10 min on ice (if it does not dissolve well it can be incubated 10–15 min at 55–60 °C).

14. Measure the concentration of the isolated RNA using Nano-Drop One and keep the samples at −70 °C until use.

3.11 RT-PCR and Sequencing Analysis to Confirm the Splicing Defect

1. For reverse transcription with the NZYRT System, use 1 μg of RNA, following the manufacturer's protocol. Random hexamers, oligo(dT) or vector-specific primer can be used. Mix RNA with NZYRT Master mix and NZYRT Enzyme mix in a final volume of 20 μL in PCR tubes, incubate 10 min at 25 °C, followed by 30 min at 50 °C, 5 min at 85 °C and cool to 4 °C.

2. Add 1 μL of NZY RNase H and incubate at 37 °C for 20 min.

3. Perform a standard PCR reaction using 1 μL cDNA and a final volume of 25 μL.

4. Run 5 μL of each sample in a 2% agarose gel.

5. Purify, quantify, and sequence the PCR product as explained above (*see* **Note 38**).

---

4 Notes

1. All the specific reagents for gene editing (crRNA, tracrRNA-ATTO550, ssDNA template, Cas9 Nuclease) explained in this protocol were obtained from IDT (Integrated DNA Technology). However, it is important to note that there are several other companies that sell these same products performing equally well. It is the researcher's decision to decide which company he wants to work with.

2. The method of delivery and/or the transfection reagent will depend on our cell line or on our preferences. Lipofection with Lipofectamine™ CRISPRMAX™ Cas9 and its Transfection

Reagent (Thermofisher) has been the method and the reagent chosen is this protocol.

3. Before starting the gene editing experiment, it is important to check that the chosen cell line expresses the gene of interest (mRNA and protein) and corresponds to a tissue relevant for your studies, i.e., splicing defect was observed in this type of cells, as splicing outcomes may depend on tissue-specific splice factors. It is also essential to take into account the organism from which the cell line is derived. For example, intronic sequences are not well conserved among species, and this is crucial when, for example, the aim is to study intronic splicing mutations.

4. It is necessary to verify the karyotype of the chosen cell line, to confirm it is normal, at least in relation to the pair of chromosomes where the gene that is going to be edited is located. Most established cell lines show aneuplodies and structural chromosomal alterations that will hinder the desired gene edition if the corresponding chromosome is affected. In our case, we tested a battery of human hepatoma cell lines, Hep3B, HepG2, Huh7 among others and selected an HepG2 cell line with two chromosomes 12 where the PAH gene is located. Karyotype analysis is a routine service offered by many human genetic diagnosis laboratories.

5. The chosen cell line should have the ability to form "single-cell colonies." This is necessary to isolate individual cells after transfection that will be subsequently expanded for genetic characterization to confirm and select correctly gene edited clones. There are different procedures for the generation of "single-cell colonies": (a) cell sorting: 1 cell/96-well-plate well using a cell sorter, (b) serial dilutions, and (c) seeding the cells at a high dilution (approximately 100 cells in one 150 mm plate).

   The election of one method or another will depend on the cell line, so it is advisable to test this before generating the colonies with the edited cells. In our hands, for example, HepG2 cells exhibited high mortality after sorting and plating in 96-well plates, so we selected option c. For some cell lines the use of conditioned medium (filtered culture medium collected from control cells) can aid the growth in the form of a colony derived from a single cell. The time of growth and appearance of single-cell colonies will depend on the type of cells you are working with. With HepG2 cells, colonies emerged and reached the correct size after circa 20 days.

6. It is advisable to have the region sequenced before starting the editing experiment to identify single-nucleotide polymorphisms in the specific cell line used which may affect the design

of RNA guides and DNA templates, as well as result in erroneous interpretation of the sequencing analysis of the edited clones (concluding there has been an extra change introduced during DNA repair after Cas9 reaction when it was already present in the sequence prior to editing).

7. There are multiple softwares for designing RNA guides for CRISPR assays. In our case we have used Breaking Cas software (http://bioinfogp.cnb.csic.es/tools/breakingcas) [19], which we find user-friendly, and the one offered by the company IDT, obtaining nearly identical results. In this sense, it is advisable to use and compare the results from at least two different softwares, to be sure that the selected guides are the most suitable.

8. It is advisable to test at least two RNA guides in a parallel and independent way. Generally, according to IDT, in 2/3 of the cases, sense sequence guides will work better than antisense guides. As we cannot predict which ones will do best for a given locus, we recommend testing both orientations.

9. As an optional step, you can pretest your RNA guides with an in vitro digestion after PCR amplification of the target region to confirm their efficiency (following IDT protocol).

10. SnapGene Viewer (https://www.snapgene.com/) has been the software used for visualization of sequences used in this project, location of crRNA, DNA templates, restriction sites, etc. and for sequence analysis of the individual edited clones. However, other programs and software can be used.

11. Cas9 nuclease cut site should be as close as possible to the sequence (nucleotide) which is to be edited. This requirement limits the region where we will design the RNA guides, especially if we want to introduce a point mutation as is the case here. It should be noted that this does not generally apply for the generation of a *knock-out* model, or in general, if we are not focused on introducing a mutation in a specific DNA position; in those cases the cut site can be in any position, so the design and choice of the RNA guide is much easier.

12. If possible, it is recommended to choose an RNA guide targeting the region that includes the nucleotide we intend to edit. Once the edition of that locus has occurred, the affinity of our RNA guide is reduced (because of a mismatch due to the mutation introduced), thus hindering possible reediting.

13. A ssDNA oligonucleotide containing the desired point mutation to be introduced is used as a template by the cell to repair the double strand break induced by Cas9 through HDR. The mutation of interest included in the ssDNA template should be in the middle of the sequence flanked by the homology arms. The length of the homology arms should be 35–40 nucleotides if it is a single-nucleotide change. Using longer homology arms

does not increase the homologous recombination success rate. However, for longer edits (e.g. insertion/deletion of several nucleotides), the length of the homology arms must also be increased.

14. For small insertions or single-nucleotide changes, ssDNA template is recommended. In other experimental situations (introduction of >100 nucleotide sequences) it may be advisable to use double stranded DNA templates.

15. In most gene editing protocols, introducing translationally silent sequence changes in the DNA template eliminating the PAM sequence is recommended, to avoid reediting of our target which may introduce unwanted changes. However, when dealing with intronic or exonic splice mutations, any extra change may alter the final splicing outcome so this should be avoided.

16. Standard desalting or HLPC are the purification methods recommended when ordering the ssDNA template. Also, especially in rich nucleases environments, phosphorothioate bonds (PS Bonds) at the extremes of the oligonucleotide are advisable, ideally putting at least two for each end of the template.

17. This protocol is written to use separate crRNA and tracrRNA. There is also the possibility of working with single guide RNA, where both are linked together, so this step will be different, refer to manufacturer's recommendations.

18. The resuspension volumes depend on the amount of purchased crRNA and tracrRNA. A table of equivalences for different quantities is available in the IDT protocols. It is important to keep in mind that the resuspended RNA oligonucleotides can be stored at −20 °C. The volumes and quantities referred to in this protocol are calculated for a 6-well plate, which has been the format used by the authors. Refer to the protocols available on the IDT website for other formats (e.g. 96-well plate).

19. The use of tracrRNA fused to the ATTO550 fluorophore is not strictly necessary but, in our hands, it was very useful for measuring transfection efficiency and to select transfected cells by fluorescence activated cell sorting (FACS) before clone generation. However, in cell types where transfection efficiency is known to be high/very high this step may be waived. In addition, there are certain cell types that are more prone to damage during the sorting process, so it would not be advisable to use this procedure to avoid increasing cell mortality. The protocol described can also be used for tracrRNA without ATTO550. In addition, it is important not to confuse transfection efficiency rate with editing efficiency, since a cell may have been transfected, but not edited. It is important to keep in mind that the success rate of the gene editing will not

only depend on the quality of the guide, but also on the transfection method, the locus we are editing, the cell type, etc.

20. The RNA Duplex can be prepared at a final concentration >1 μM and stored at −20 °C during, at least, 6 months. Before use, it should be diluted in Nuclease-Free Duplex Buffer to a working concentration of 1 μM.

21. IDT provides Cas9 nuclease at a stock concentration of 62 μM. It can be diluted in different buffers, such as PBS or Cas9 Working Buffer (20 mM HEPES, 150 mM KCl, pH 7.5). This will depend on our cell type. In our case we have used OptiMEM to dilute the Cas9 enzyme. It will be important to take these details into account when purchasing Cas9 nuclease from other companies.

22. The final concentration of the ssDNA template is variable depending on the cell type, delivery method, etc. In this case (transfection of HepG2 cells with Lipofectamine (CRISPR-Max)), a final concentration of 3 nM ssDNA template was used, following the manufacturer's recommendations. Transfecting higher amounts of ssDNA template does not ensure a higher rate of editing success. In addition, large amounts of DNA oligonucleotide can become toxic for the cells and increase cell mortality.

23. In initial experiments, it is advisable to perform the reverse transfection of each crRNA guide in triplicate (three 6-well plate wells/crRNA).

24. As with any transfection assay, it is advisable to split and pass the cells at least once after defrosting before starting the test.

25. Before generating colonies derived from a single cell, it is important to freeze the remaining total pool of cells transfected with each crRNA. In the event of any problem we could defrost those cells to generate the colonies again without the need to repeat the transfection.

26. Once the colonies have grown to a size allowing us to handle them efficiently, they must be expanded for analysis. You can select as many colonies as you can manage. You must consider that expansion, cultivation, and analysis of individual colonies require considerable effort and dedication. Normally we grow around 50–70 colonies for each crRNA used.

27. The system used to select colonies and pick them can be very variable. For example, cloning cylinders can be used or other methods of choice of the researcher.

28. We expanded the single-cell colonies in 24-well plates, but this can be modified according to the researcher's preferences and/or cell line characteristics using plates with different formats. In our case, once the cells are confluent, we divide each

well of the 24-well plate into two wells of a 12-well plate. It is important to keep accurate record of each duplicate, since one of them will be used to extract DNA for analysis, and the other will be used to freeze the colony and, in case it is the one selected, expand it for further characterization and use.

29. The analysis of the colonies derived from a single cell is necessary to identify edited ones. In our case, the point mutation that we are introducing generates a new restriction site for the PshAI enzyme. This is very useful to rapidly and easily screen by RFLP analysis for the presence of the introduced mutation, although the edited region must be verified by sequence analysis. In some applications, translationally silent changes are introduced in the donor template near the mutation to create/destroy a restriction site, thus allowing RFLP screening. However, this is not recommended for splicing mutations as any nearby change may alter the splicing outcome. Alternative approaches to evaluate edition efficiency include next-generation sequencing approaches or digital droplet PCR.

30. Other commercial kits or in-house methods can be used for DNA extraction.

31. Other alternative software and resources can be used with the same objective. Primers are designed to amplify the region with the desired change, which should ideally be in the middle of the amplicon, so after digesting with the corresponding enzyme and running the products in an agarose gel we can easily distinguish digested and undigested DNA bands, which will facilitate the identification of the positive clone. Care should be taken during primer design to ensure that there are no other restriction sites for the corresponding enzyme (in our case PshAI) within the amplicon.

32. Due to the high number of colonies, it is very laborious to analyze all of them individually. Therefore, it is advisable to make pools with DNA extracted from 4 or 5 colonies, mixing them to obtain 200 ng of total DNA. Once edition is observed in the RFLP analysis, colonies will then be analyzed individually.

33. It is not necessary to purify the PCR products before restriction enzyme digestion. Purification of PCR products does not improve digestion efficiency, as the PCR product is diluted enough so that the different components of the PCR reaction do not interfere with the enzymatic activity.

34. The conditions, temperatures, and times of the restriction reaction may depend on the enzyme and/or the trademark.

35. Usually, amplification and subsequent sequencing of the three possible off-targets with the highest scores identified by the software used is enough. Based on our experience we can

conclude that off-targets, although it is important to sequence and validate them, are not the biggest problem. However, we frequently found extra changes in the area near the edited nucleotide (on-target). In this sense, these errors have been the main problem and the cause of having had to discard many clones before finding the final positive one.

36. There are many commercially available kits for purification of PCR products. We routinely use Cycle Pure Kit (Omega).

37. Sequencing is necessary for the validation of the positive clone. And to discard off-target effects in the correctly edited clone. We should confirm that no extra changes have been made in the edited region. IMPORTANT: Do not confuse these random changes that CRISPR introduces when repairing the DSB in the DNA (on-target effects) with potential off-targets, which are locus to which our crRNAs can bind and induce a DSB in the DNA.

38. It is important, once the positive clone is selected and genetically analyzed, to carry out the phenotypic characterization as cellular model of the disease phenotype, to confirm that it accurately recapitulates the splicing defect, resulting (in our case) in the absence of protein and activity. To that aim, RT-PCR and cDNA sequencing, followed by Western blot analysis of PAH protein and PAH activity assay were performed. The specific analyses to be performed will depend on each case according to the aim of the study, but, in the case of splicing mutations they should include at least RT-PCR and subsequent cDNA sequencing analysis.

References

1. Montes M, Sanford BL, Comiskey DF, Chandler DS (2019) RNA splicing and disease: animal models to therapies. Trends Genet 35(1): 68–87. https://doi.org/10.1016/j.tig.2018.10.002

2. Lim KH, Ferraris L, Filloux ME, Raphael BJ, Fairbrother WG (2011) Using positional distribution to identify splicing elements and predict pre-mRNA processing defects in human genes. Proc Natl Acad Sci U S A 108(27): 11093–11098. https://doi.org/10.1073/pnas.1101135108

3. Sterne-Weiler T, Howard J, Mort M, Cooper DN, Sanford JR (2011) Loss of exon identity is a common mechanism of human inherited disease. Genome Res 21(10):1563–1571. https://doi.org/10.1101/gr.118638.110

4. Scotti MM, Swanson MS (2016) RNA mis-splicing in disease. Nat Rev Genet 17(1): 19–32. https://doi.org/10.1038/nrg.2015.3

5. Aartsma-Rus A (2016) New momentum for the field of oligonucleotide therapeutics. Mol Ther 24(2):193–194. https://doi.org/10.1038/mt.2016.14

6. Kim J, Hu C, Moufawad El Achkar C, Black LE, Douville J, Larson A, Pendergast MK, Goldkind SF, Lee EA, Kuniholm A, Soucy A, Vaze J, Belur NR, Fredriksen K, Stojkovska I, Tsytsykova A, Armant M, DiDonato RL, Choi J, Cornelissen L, Pereira LM, Augustine EF, Genetti CA, Dies K, Barton B, Williams L, Goodlett BD, Riley BL, Pasternak A, Berry ER, Pflock KA, Chu S, Reed C, Tyndall K, Agrawal PB, Beggs AH, Grant PE, Urion DK, Snyder RO, Waisbren SE, Poduri A, Park PJ, Patterson A, Biffi A, Mazzulli JR, Bodamer O, Berde CB, Yu TW (2019) Patient-customized oligonucleotide therapy for a rare genetic disease. N Engl J Med 381(17):1644–1652. https://doi.org/10.1056/NEJMoa1813279

7. Aartsma-Rus A, Corey DR (2020) The 10th oligonucleotide therapy approved: golodirsen for Duchenne muscular dystrophy. Nucleic Acid Ther 30(2):67–70. https://doi.org/10.1089/nat.2020.0845

8. Kemaladewi DU, Maino E, Hyatt E, Hou H, Ding M, Place KM, Zhu X, Bassi P, Baghestani Z, Deshwar AG, Merico D, Xiong HY, Frey BJ, Wilson MD, Ivakine EA, Cohn RD (2017) Correction of a splicing defect in a mouse model of congenital muscular dystrophy type 1A using a homology-directed-repair-independent mechanism. Nat Med 23(8):984–989. https://doi.org/10.1038/nm.4367

9. Schneller JL, Lee CM, Bao G, Venditti CP (2017) Genome editing for inborn errors of metabolism: advancing towards the clinic. BMC Med 15(1):43. https://doi.org/10.1186/s12916-017-0798-4

10. Bollen Y, Post J, Koo BK, Snippert HJG (2018) How to create state-of-the-art genetic model systems: strategies for optimal CRISPR-mediated genome editing. Nucleic Acids Res 46(13):6435–6454. https://doi.org/10.1093/nar/gky571

11. Maule G, Casini A, Montagna C, Ramalho AS, De Boeck K, Debyser Z, Carlon MS, Petris G, Cereseto A (2019) Allele specific repair of splicing mutations in cystic fibrosis through AsCas12a genome editing. Nat Commun 10(1):3556. https://doi.org/10.1038/s41467-019-11454-9

12. Xu S, Luk K, Yao Q, Shen AH, Zeng J, Wu Y, Luo HY, Brendel C, Pinello L, Chui DHK, Wolfe SA, Bauer DE (2019) Editing aberrant splice sites efficiently restores beta-globin expression in beta-thalassemia. Blood 133(21):2255–2262. https://doi.org/10.1182/blood-2019-01-895094

13. Doudna JA (2020) The promise and challenge of therapeutic genome editing. Nature 578(7794):229–236. https://doi.org/10.1038/s41586-020-1978-5

14. Mojica FJ, Diez-Villasenor C, Garcia-Martinez J, Soria E (2005) Intervening sequences of regularly spaced prokaryotic repeats derive from foreign genetic elements. J Mol Evol 60(2):174–182. https://doi.org/10.1007/s00239-004-0046-3

15. Jinek M, Chylinski K, Fonfara I, Hauer M, Doudna JA, Charpentier E (2012) A programmable dual-RNA-guided DNA endonuclease in adaptive bacterial immunity. Science 337(6096):816–821. https://doi.org/10.1126/science.1225829

16. Knott GJ, Doudna JA (2018) CRISPR-Cas guides the future of genetic engineering. Science 361(6405):866–869. https://doi.org/10.1126/science.aat5011

17. Morava E, Rahman S, Peters V, Baumgartner MR, Patterson M, Zschocke J (2015) Quo vadis: the re-definition of "inborn metabolic diseases". J Inherit Metab Dis 38(6):1003–1006. https://doi.org/10.1007/s10545-015-9893-x

18. Martinez-Pizarro A, Dembic M, Perez B, Andresen BS, Desviat LR (2018) Intronic PAH gene mutations cause a splicing defect by a novel mechanism involving U1snRNP binding downstream of the 5′ splice site. PLoS Genet 14(4):e1007360. https://doi.org/10.1371/journal.pgen.1007360

19. Oliveros JC, Franch M, Tabas-Madrid D, San-Leon D, Montoliu L, Cubas P, Pazos F (2016) Breaking-Cas-interactive design of guide RNAs for CRISPR-Cas experiments for ENSEMBL genomes. Nucleic Acids Res 44 (W1):W267–W271. https://doi.org/10.1093/nar/gkw407

5

Establishment of In Vitro Brain Models for AON Delivery

Elena Daoutsali and Ronald A. M. Buijsen

Abstract

Progress in stem cell biology has made it possible to generate human-induced pluripotent stem cells (hiPSC) that can be differentiated into complex, three-dimensional structures, where the cells are spatially organized. To study brain development, Lancaster and colleagues developed an hiPSC-derived three-dimensional organoid culture system, termed cerebral organoids, that develop various discrete, although interdependent, brain regions. Here we describe in detail the generation of cerebral organoids using a modified version of the culture protocol.

Key words Cerebral organoid, Disease modeling, Induced pluripotent stem cells

1   Introduction

Many brain disorders are hereditary diseases with a known genetic cause, which allowed scientists to generate animal models to study disease progression, understand disease mechanisms, and perform therapeutic intervention studies [1, 2]. However, (1) mice are different from humans, and it is difficult to translate results from animal experiments into clinical application; (2) the genetic cause of many diseases is not yet known; (3) many disease-causing genes are mainly expressed in the cells that are affected; (4) for many of them, there are no (humanized-)mouse models available; (5) there is governmental and public pressure to advance the development of alternative model systems to replace animal studies. This emphasizes the need for patient-derived disease models that bridge the translational gap between animal models and human clinical trials. Progress in stem cell biology has made it possible to generate human induced pluripotent stem cells (hiPSCs) [3] that can be differentiated into the important cell types of the brain, neurons, and astrocytes [4, 5]. The disadvantage of these 2D models is that they are descriptive at a cellular level, but they fail to adequately provide the details that could be derived from a more complex, three-dimensional structure, where the cells are spatially organized

[6]. In 2013, Lancaster and colleagues developed a hiPSC-derived three-dimensional organoid culture system, termed cerebral organoids, that develop various discrete, although interdependent, brain regions [7]. These organoids recapitulate many features of human cortical development, including a progenitor zone organization with abundant outer radial glial stem cells [8].

Here we describe the generation of cerebral organoids using a modified version of the Lancaster protocol [7, 9]. In short, feeder-free cultured hiPSCs were dissociated and replated in neural induction medium in a non-adherent cell culture plate, and differentiated for 100 days (Fig. 1). Cryosections of these organoids can be used for immunofluorescence studies. Organoids can be used for many different purposes including disease modeling, studying disease mechanisms, or analyzing therapeutic interventions (using for example antisense oligonucleotides) at any given time point.

2 Materials

2.1 Neuroectodermal Differentiation

1. mTeSR™1.
2. Matrigel.
3. Dulbecco's Modified Eagle's Medium (DMEM) and Ham's F-12 Nutrient Mixture (DMEM/F12).
4. ACCUTASE™.
5. STEMdiff™ Neural Induction Medium (NIM).
6. Y-27632.
7. v-bottom shape 96-well plate.

2.2 Neurospheres

1. Neurosphere medium: DMEM/F12 and Neurobasal medium 1:1, 1:200 N2 supplement, 1:100 B-27 supplement (without vitamin A), 1:100 L-glutamine, 0.05 mM non-essential amino acids (MEM-NEAA), 100 U/ml penicillin, 100 μg/ml streptomycin, 1.6 mg/l insulin, 0.05 mM β-mercaptoethanol.
2. Wide orifice pipette tips.
3. Organoid embedding sheet (or parafilm and a 200 μl tip box).

2.3 Organoids

1. Brain organoid medium: DMEM/F12 and Neurobasal medium 1:1, 1:200 N_2 supplement, 1:100 B27 supplement w/o vitamin A, 1:100 L-glutamine, 0.05 mM MEM-NEAA, 100 U/ml penicillin, 100 μg/ml streptomycin, 1.6 mg/l insulin, 0.5 μM dorsomorphin, 5 μM SB431542, 0.05 mM β-mercaptoethanol.
2. Spinner flask or 6-well plates.
3. Bioreactor or shaker.

Initiation of neuroectoderm		Neurosphere formation	Cerebral organoid maturation		
Day 3	Day 5	Day 10	Day 16	Day 23	Day 40

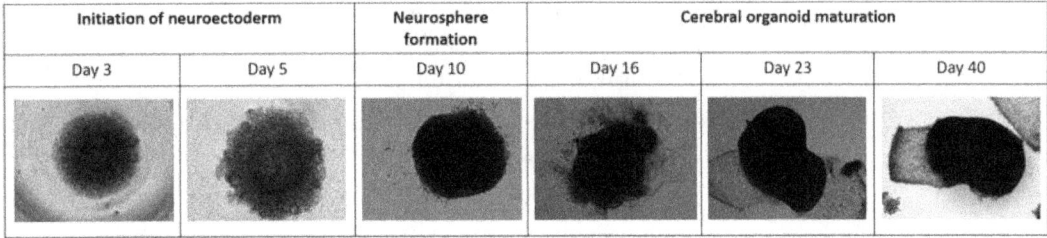

Fig. 1 Cerebral organoids during the various stages of organoid culturing. Organoids are cultured using a modified version of the Lancaster protocol. After 5 days of neuroectodermal differentiation, the neurospheres are embedded in Matrigel and cultured in the neurosphere medium in a 6-well plate for 5 days. For cerebral organoid maturation, the embedded neurospheres are transferred into a spinner flask and can be used for downstream applications if needed

2.4 Fixation and Embedding

1. Wide orifice pipette tips.
2. Dulbecco's Phosphate Buffered Saline (DPBS).
3. 4% paraformaldehyde (PFA).
4. 30% sucrose in distilled water.
5. Embedding mold.
6. Optimum cutting temperature compound (OCT).

2.5 Cryosectioning and Immunofluorescent Staining

1. PLL-coated glass cryoslides.
2. Barrier pen.
3. PBS-glycine: 200 mM of glycine in DPBS.
4. Blocking Solution: 5% goat or horse serum, 0.1% Triton X-100, 200 mM glycine in Dulbecco's Phosphate-Buffered Saline.
5. Immunobuffer solution: 1% goat or horse serum, 0.1% Triton X-100 in Dulbecco's Phosphate-Buffered Saline.
6. Prolong Diamond Antifade Mounting (+DAPI).

3 Methods

3.1 Neuroectodermal Differentiation

1. Culture hiPSCs under feeder-free conditions in mTeSR™1 in a culture dish coated with Matrigel. For neuroectodermal differentiation one 100 mm cell culture dish is required (*see* **Note 1**).
2. When the hiPSCs are ready for passaging, wash the hiPSCs with 10 ml pre-warmed (37 °C) DMEM/F12.
3. Remove DMEM/F12, add 2 ml pre-warmed ACCUTASE™ and incubate for 5 min at 37 °C and 5% CO_2, allowing cells to detach (*see* **Note 2**).

4. Pipette the cell suspension up and down 3–5 times using a 1-ml micropipette to make a single-cell suspension and collect the suspension in a 15-ml tube (*see* **Note 3**).

5. Add 6 ml of pre-warmed DMEM/F12, wash the culture dish and collect the suspension in the same 15-ml tube as in **step 4**.

6. Centrifuge at $300 \times g$ for 5 min at room temperature.

7. Remove supernatant and resuspend the cell pellet in 1 ml of NIM supplemented with 10 μM Y-27632 and count the cells using a cell counter.

8. Dilute the cell suspension in NIM supplemented with 10 μM Y-27632 to 4.5×10^5 cells per ml and add 100 μl per well in a non-adherent, v-bottom shape, 96-well plate (*see* **Note 4**).

9. Centrifuge the plate at $500 \times g$ for 3 min at room temperature and incubate at 37 °C and 5% CO_2.

10. Change medium by carefully removing 50 μl NIM from the top of the wells, without disturbing the embryoid bodies, and by adding 50 μl of fresh NIM. Medium changes should be done daily for the next 5 days (*see* **Note 5**).

3.2 Neurosphere Embedding

1. Use an 1-ml micropipette with a wide orifice pipet tip to place the neurospheres on a silicone organoid embedding sheet (*see* **Notes 6** and **7**).

2. Carefully remove the medium from the well, without disturbing the neurosphere, and expel it back to dislodge the neurosphere from the bottom of the well.

3. Use a wide orifice tip to collect the neurosphere and transfer it to the embedding sheet.

4. Carefully remove as much liquid from the embedding sheet as possible.

5. Use an ice-cold tip to add a drop of Matrigel onto each neurosphere.

6. Use an ice-cold tip to place the neurosphere in the center of the Matrigel droplet.

7. Incubate at 37 °C and 5% CO_2 for 15 min.

8. Carefully wash the neurospheres from the embedding sheet into a 100-mm culture dish by flushing them with the neurosphere medium and incubate at 37 °C and 5% CO_2. The total volume of neurosphere medium in the dish is 10 ml (*see* **Note 8**).

9. Add 2 ml of fresh neurosphere medium on day 2.

10. Use the embedded neurospheres 4 days after the embedding for the next step in the protocol.

3.3 Organoids

1. Carefully transfer the embedded neurospheres into a spinner flask containing 100 ml of pre-warmed organoid medium using a 2-ml serological pipette (*see* **Note 9**).

2. Place the spinner flask on the magnetic stirring platform in the incubator and use a stirring program at 25 rpm.

3. Culture the organoids on the magnetic stirring platform at 37 °C and 5% CO_2 for up to 100 days (*see* **Notes 10** and **11**).

4. Change the medium weekly by removing 50 ml of the organoid medium and adding 50 ml of fresh organoid medium (*see* **Note 12**). To refresh the medium, remove the spinner flask and let the organoids sink to the bottom for 5 min. Then carefully remove the medium, without disturbing the organoids.

3.4 Fixation and Embedding

1. Collect the organoids with an 1-ml micropipette with a wide orifice pipet tip and transfer them in a 60-mm dish or in a 6-well plate (*see* **Note 6**).

2. Wash the organoids with 5 ml of prewarmed Dulbecco's Phosphate-Buffered Saline.

3. Use a wide orifice tip to transfer each organoid separately into a 1.5-ml Eppendorf tube with 500 μl of 4% PFA. Incubate the organoids for 30 min at room temperature (*see* **Notes 13** and **14**).

4. Remove the 4% PFA solution and wash the organoids twice with 1 ml DPBS for 5 min (*see* **Note 15**).

5. Remove DPBS and add 1 ml of 30% sucrose in distilled water per tube to dehydrate the organoids and incubate the organoids at 4 °C overnight (*see* **Note 16**).

6. Fill a Peel-A-Way embedding molds with 400 μl of optimum cutting temperature compound (until the middle) and use an inoculation loop to place the organoid in the center of the mold. Label the rim of the mold with the sample name (*see* **Note 17**).

7. Snap-freeze the organoid-containing mold with ethanol on dry-ice and store at −80 °C until further use.

3.5 Cryosectioning and Immunofluorescent Staining

1. Section cryoprotected frozen organoids into 16- to 20-μm-thick slices on PLL-coated glass cryoslides using a cryostat (*see* **Notes 18** and **19**).

2. Thaw (if they were frozen) and dry the slides for 30 min at room temperature.

3. Draw a hydrophobic barrier around each section using a barrier pen.

4. To quench the PFA-induced autofluorescence wash the slides twice with 200 μL of PBS-glycine for 3 min.

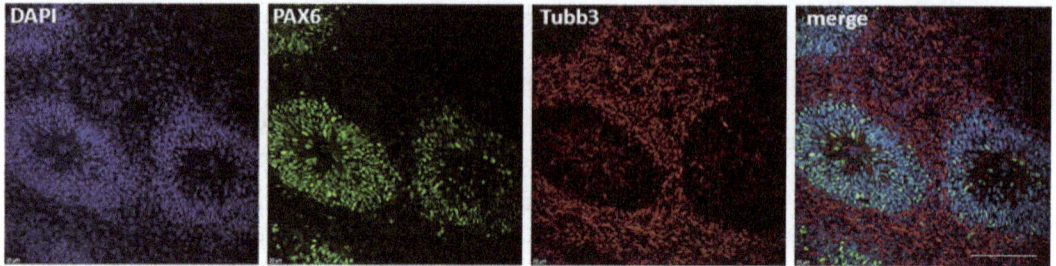

Fig. 2 Immunofluorescent staining of a cortical plate structure. Cortical plate structure in cerebral organoids stained with DAPI (blue), the neural progenitor marker PAX6 (green), and the neural marker TUBB3 (red). The scale bar represents 100 μm

5. Block nonspecific binding by adding 100 μl of Blocking Solution to the section for 1 h at room temperature.

6. Add 100 μl of primary antibody diluted in immunobuffer solution and incubate the slides overnight at 4 °C (*see* **Note 20**).

7. Wash three times with DPBS for 5 min.

8. Incubate the sections with the secondary antibody in immunobuffer solution for 1.5–2 h at room temperature (*see* **Note 21**).

9. Wash the slides three times in PBS.

10. Put a drop of Prolong Diamond Antifade Mountant (+DAPI) on the section and put a coverslip on top. Leave overnight at room temperature.

11. Store in fridge at 4 °C until performing microscopy. An example of a cortical plate structure in cerebral organoids stained with the neural progenitor marker PAX6 and the neural marker TUBB3 can be seen in Fig. 2.

4   Notes

1. Change mTeSR™1 daily and passage hiPSCs after 5–7 days. hiPSCs are ready to passage when the majority of the colonies are large, compact, and have centers that are dense compared to their edges. Only use undifferentiated, high-quality hiPSCs that are fully characterized according to the latest human pluripotent stem cell registry guidelines (https://hpscreg.eu/).

2. 1 ml ACCUTASE™ per 25 cm^2 surface area.

3. ACCUTASE™ is a cell detachment solution of proteolytic and collagenolytic enzymes and does not need to be neutralized.

4. Make sure that the cells are equally distributed in the suspension by mixing the tube regularly.

5. There is more evaporation from the four corner wells. Do not use these wells or add extra medium daily (up to 100 µl total culture volume).

6. The wide orifice tip can be replaced by a cut tip. The cut on the tip should be done by using sterile scissors. You can effectively sterilize scissors in an autoclave, but tools can also be sterilized in alcohol or a flame.

7. The silicone organoid embedding sheet can be replaced by placing parafilm over an empty 200-µl tip box and using a finger to make small holes.

8. A maximum of 20 neurospheres per 100-mm cell culture dish.

9. A maximum of 20 organoids per culture flask.

10. When spinner flasks are not available, non-adhesive culture plates on a rotating platform can be used. Use a shaking program of 75 rpm.

11. If an orbital shaker and a 6-well plate are used, do not transfer more than 10 organoids per well. Moreover, monitor the organoids regularly to reduce the chance of them sticking together. If organoids stick together, you can easily separate these by using a sterile pipet tip as a knife.

12. Medium changes need to be done more often when there is a color change of the medium. Check the organoids regularly and replace the medium as done in **step 3** in Subheading 3.3 when needed. For a 6-well plate, a volume of 3-ml medium per well can be used and medium should be changed twice a week and/or when there is a color change of the medium.

13. For a big batch of organoids, fixation can be done in a 6-well plate.

14. For larger organoids (>2 mm), a 4% PFA incubation at 4 °C overnight is recommended.

15. You can store the organoids in 1 ml DPBS at 4 °C for up to 7 days.

16. After the addition of 30% sucrose solution, the organoids should float at the surface, and by the next day, the organoids should sink down to the bottom of the tube. You can keep the organoid in 30% sucrose for up to a month. The recommended time for organoids larger than 4 mm is 5–7 days.

17. To better visualize the organoid while cryosectioning, add Trypan-blue diluted 1:50 in DPBS before embedding the organoids in OCT (and after 30% sucrose) for 15 min at room temperature. The outer area of the organoid will be colored blue.

18. Up to 100 sections can be obtained from one organoid.

19. The slides can be stored at −80 °C.

20. Put a wet tissue inside the box to prevent the slides from drying out.

21. Keep in the dark from here on.

References

1. Chesselet MF, Carmichael ST (2012) Animal models of neurological disorders. Neurotherapeutics 9(2):241–244. https://doi.org/10.1007/s13311-012-0118-9

2. Berman RF, Buijsen RA, Usdin K, Pintado E, Kooy F, Pretto D, Pessah IN, Nelson DL, Zalewski Z, Charlet-Bergeurand N, Willemsen R, Hukema RK (2014) Mouse models of the fragile X premutation and fragile X-associated tremor/ataxia syndrome. J Neurodev Disord 6(1):25. https://doi.org/10.1186/1866-1955-6-25

3. Takahashi K, Tanabe K, Ohnuki M, Narita M, Ichisaka T, Tomoda K, Yamanaka S (2007) Induction of pluripotent stem cells from adult human fibroblasts by defined factors. Cell 131(5):861–872. https://doi.org/10.1016/j.cell.2007.11.019

4. Hu BY, Weick JP, Yu J, Ma LX, Zhang XQ, Thomson JA, Zhang SC (2010) Neural differentiation of human induced pluripotent stem cells follows developmental principles but with variable potency. Proc Natl Acad Sci U S A 107(9):4335–4340. https://doi.org/10.1073/pnas.0910012107

5. Shi Y, Kirwan P, Livesey FJ (2012) Directed differentiation of human pluripotent stem cells to cerebral cortex neurons and neural networks. Nat Protoc 7(10):1836–1846. https://doi.org/10.1038/nprot.2012.116

6. Centeno EGZ, Cimarosti H, Bithell A (2018) 2D versus 3D human induced pluripotent stem cell-derived cultures for neurodegenerative disease modelling. Mol Neurodegener 13(1):27. https://doi.org/10.1186/s13024-018-0258-4

7. Lancaster MA, Knoblich JA (2014) Generation of cerebral organoids from human pluripotent stem cells. Nat Protoc 9(10):2329–2340. https://doi.org/10.1038/nprot.2014.158

8. Lancaster MA, Renner M, Martin CA, Wenzel D, Bicknell LS, Hurles ME, Homfray T, Penninger JM, Jackson AP, Knoblich JA (2013) Cerebral organoids model human brain development and microcephaly. Nature 501(7467):373–379. https://doi.org/10.1038/nature12517

9. Gabriel E, Gopalakrishnan J (2017) Generation of iPSC-derived human brain organoids to model early neurodevelopmental disorders. J Vis Exp 122:55372. https://doi.org/10.3791/55372

6

Antisense RNA Therapeutics: A Brief Overview

Virginia Arechavala-Gomeza ⑩ and Alejandro Garanto ⑩

Abstract

Nucleic acid therapeutics is a growing field aiming to treat human conditions that has gained special attention due to the successful development of mRNA vaccines against SARS-CoV-2. Another type of nucleic acid therapeutics is antisense oligonucleotides, versatile tools that can be used in multiple ways to target pre-mRNA and mRNA. While some years ago these molecules were just considered a useful research tool and a curiosity in the clinical market, this has rapidly changed. These molecules are promising strategies for personalized treatments for rare genetic diseases and they are in development for very common disorders too. In this chapter, we provide a brief description of the different mechanisms of action of these RNA therapeutic molecules, with clear examples at preclinical and clinical stages.

Key words RNA therapy, Antisense oligonucleotides, Clinical trials, Splicing, Personalized medicine

1 Introduction

Nucleic acid therapeutics is still a growing field. With the irruption of the mRNA vaccines against SARS-CoV-2 special attention has been given to this type of therapies but other types of nucleic acid therapeutics, coined antisense oligonucleotides (AONs), have been studied for many years. Although only a dozen therapeutic oligonucleotides have been formally approved for clinical use, there are many new such drugs in the pipeline for a plethora of (mainly rare) diseases. These AON molecules interact with different nucleic acids (mRNA, non-coding RNA, and DNA) thanks to sequence specific Watson–Crick base pairing. Their mechanism of action, that may be designed to bind specific targets, makes these drugs easy to design, less likely to cause side effects and, therefore, potential candidates to lead the next wave of precision medicine. In this chapter, we describe the most frequently used AON-based therapeutic strategies, their mechanisms of action (Fig. 1), and the results of several clinical trials, with special emphasis in eye and muscle diseases.

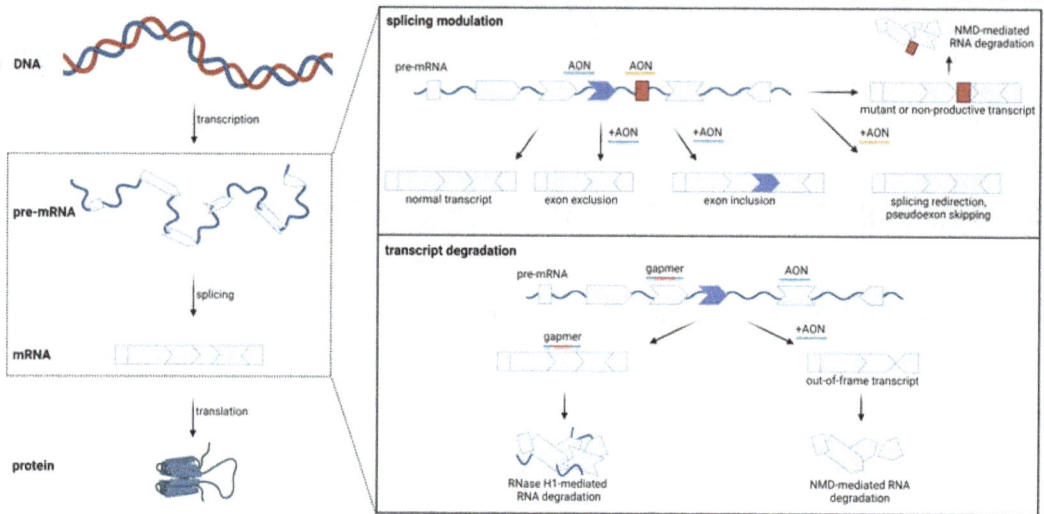

Fig. 1 Schematic representation of the multiple mechanisms of action of antisense oligonucleotide (AON) molecules. AONs can act at pre- and mRNA levels of the synthesis of a functional protein (left panel). They can be used to modulate splicing (upper right panel) or to degrade (pre-)mRNA (lower right panel). Splice-modulating AONs bind to pre-mRNA and promote the insertion or skipping of regular exons. In addition, they can redirect splicing when mutations in a gene lead to splicing defects (such as pseudoexon insertions). This splicing modulation causes the degradation of the transcript and a consequent reduction of protein levels. Alternatively, transcript degradation can also be achieved by using AONs binding to the pre-mRNA to disrupt the open reading frame and degrade transcripts via nonsense-mediated decay (NMD). Gapmers, in contrast, can bind to both pre-mRNA and mRNA and activate RNase-H1 RNA degradation. (Created with BioRender. com)

2   Mechanisms of Action

2.1   Splicing Modulation

The majority of existing therapeutic AONs are designed to alter the splicing pattern of specific pre-mRNAs [1]. This can be used to treat disorders caused by splicing alterations, which with the current widespread availability of better and cheaper sequencing options are being identified more easily and can be specifically targeted [2].

2.1.1   Exon Exclusion (Shortened Proteins)

In many genes, deleting an exon would result in the production of a non-functional protein, as their structure and function would be compromised. However, there are some cases in which internally "trimmed" proteins could be partially functional due to the existence of less-vital structures within a large protein. Mutations in the *DMD* gene disrupt the open reading frame (ORF) and the expression of the dystrophin protein, leading to Duchene muscular dystrophy (DMD). In contrast, in the much milder Becker muscular dystrophy mutations in the same gene that maintain the ORF produce an internally deleted but functional protein. This real-life example was rapidly seized as an opportunity to achieve the same effect using therapeutic AONs, and several of the recently approved AON molecules target different *DMD* exons [3–9]. Because there are many different *DMD* mutations, the skipping specific exons would be therapeutic for different subsets of patients.

This concept has also been employed in the development of new AONs to skip in-frame exons carrying single-nucleotide changes generating premature stop codons in large genes. Mutations in *USH2A* cause either Usher syndrome (deafness combined with blindness) or isolated blindness in the form of retinitis pigmentosa. Exon 13 of *USH2A* is prone to carry truncating variants and by deleting it, a protein with residual function is potentially produced [10, 11]. This is also the case of a stop codon introduced by a variant in exon 36 of *CEP290*, which is naturally skipped at low levels in the retina and involved in retinal dystrophy. AONs designed to skip exon 36 restored the reading frame and produced a functional protein able to rescue the cellular phenotype in patient-derived cells [12]. Following the same strategy, AON molecules to skip different exons of *COL7A1* have been developed for dystrophic epidermolysis bullosa, a skin disease inherited in both dominant and recessive fashion [13–16].

2.1.2 Exon Inclusion

A seemingly opposite mechanism of action is at the core of nusinersen, an AON approved for the treatment of spinal muscular atrophy (SMA). In this case, mutations in the *SMN1* gene cause low or lack of SMN protein production. However, SMN protein can be produced by two nearly identical genes, *SMN1* and *SMN2*. The latter, however, contributes at very low levels due to the high rate of exon-7 skipping that disrupts the ORF. Nusinersen is used to alter the splicing of *SMN2* and include exon 7, and therefore produce sufficient amounts of SMN protein to ameliorate the patient's disease [17, 18].

2.1.3 Splicing Redirection

Variants close to the splice sites result often in either exon skipping or exon elongation. In the second scenario, mutations decrease the recognition of the original splice site and a cryptic splice site present in the intron is recognized; in the most extreme case, the entire intron is retained. Exonic variants may cause a synonymous or a predicted non-deleterious missense change at protein level, and in addition they can have a dramatic effect at RNA level by creating a novel splice site. In any case, these splicing defects are also amenable for AON intervention. For instance, a mutation in exon 3 of *USH1C* linked to deafness generates a novel splice donor site (SDS) upstream the regular SDS of the exon. This new SDS is preferentially used by the cells, leading to a disrupted reading frame. By using AONs to block the newly generated SDS, the normal transcript can be produced [19]. A similar approach has been used to target the exon elongation caused by near-exon intronic variants in *ABCA4* linked to Stargardt macular degeneration. This study showed that by blocking the newly created SDS the normal splicing can be restored. However, this approach turned out to be not that successful when targeting exon elongations caused by variants in the splice acceptor site (SAS) [20].

Another elegant way to modulate splicing using AONs is by targeting the non-productive transcripts. These transcripts often are generated by (a) alternative splicing causing insertion or skipping of exons; (b) using alternative cryptic splice sites; and (c) retaining the introns. In any case, these splicing events lead to a disrupted ORF being the transcript degraded via nonsense-mediated decay (NMD). A very recent study has shown that 1246 potentially disease-associated genes present at least one of these non-productive transcripts. By targeting these splicing events to insert or skip an exon, exclude a retained intron, or redirect splicing when a cryptic splice site is used, the overall protein levels can be increased and this might be a promising therapeutic tool for haploinsufficiency cases [21].

2.1.4 Deep-Intronic Variants

For many years intronic variants have been ignored. This is mainly because they lay in the non-informative regions of our genome, the introns. However, the implementation of novel and more robust sequencing tools has contributed to solve the missing heritability in several diseases by discovering deep-intronic mutations with a detrimental effect on pre-mRNA splicing. The study of these variants is complex but with the help of novel bioinformatic (e.g. SpliceAI [22]) and molecular tools (mini/midi/maxi-genes [23]), it is possible to predict the effect at pre-mRNA level. These variants often result in the insertion of a pseudoexon, a piece of the intron that it is recognized as an exon and leads to a disruption of the ORF and consequently the generation of a premature stop codon. Pseudoexon exclusion can be achieved by using AONs. In the *ABCA4* gene, AONs have shown splicing correction for most of the 35 intronic variants identified as cause of Stargardt disease [24–30].

2.2 Transcript Degradation

Antisense technology can be extremely useful to degrade transcripts and cause gene silencing (knockdown). From the therapeutic perspective, this might be a potential tool to treat autosomal dominant diseases caused by dominant-negative mutations. In this case, by degrading specifically the mutant allele, the correct protein can perform its function properly.

2.2.1 RNase H1-Activating Antisense Oligonucleotides (Gapmers)

These antisense molecules are characterized for being able to actively reduce the levels of the mRNAs in the nucleus and cytoplasm [31], therefore they are very useful to downregulate gene expression. These RNase H1-activating AONs or gapmers are chimeric molecules linked using a phosphorothioate (PS) backbone that usually present a conformation 5-10-5, where the two arms consist of five modified RNA nucleotides ($2'$-O-methoxyethyl ($2'$MOE), $2'$-O-methyl ($2'$OMe) or locked nucleic acid (LNA)) flanking ten DNA nucleotides [32]. The first-ever AON approved by the FDA was fomivirsen, a first-generation RNase H1-activating AON [33–36] (*see* Subheading 3). However, this is the only RNase H1-activating AON that does not have the chimeric RNA/DNA structure. So far, four molecules using this mechanism of action have received FDA and/or EMA approval to treat different disease conditions [32].

Gapmers can be used to downregulate genes in allele-independent and allele-specific manner. Below, we review some examples of each case.

Allele-independent mRNA degradation is often used to target genes or pathways that are overexpressed in certain disease conditions or can worsen the disease progression. Thus, reducing the levels of particular genes can be very beneficial. This is the case for two of the approved AON drugs: mipomersen and volanesorsen. These molecules target the mRNA of the apolipoprotein B-100 in familial hypercholesterolemia or apolipoprotein C3 in familial chylomicronaemia syndrome, hypertriglyceridemia and familial partial lipodystrophy, respectively, to lower the levels of specific lipids increased in these diseases [37–41].

In contrast, allele-specific mRNA degradation aims to target only the mutant allele. This way, specific mutations that cause a dominant-negative effect can be targeted. This is the case of inotersen, a gapmer designed to target the mRNA encoding the transthyretin (TTR) protein in autosomal dominant hereditary transthyretin amyloidosis [42, 43]. A single-nucleotide change in the gene produces misfolding of the TTR protein. As TTR protein needs to tetramerize in order to conduct its function, the addition of mutant monomers into the tetramer affects the overall function. Systemic amyloid depositions are formed, leading to progressive polyneuropathy of the sensory and motor systems with multiorgan dysfunction in late-disease stages. The therapeutic gapmer targets the mutant allele to reduce the amount of tetramers containing the mutant protein, and therefore prevent the aforementioned depositions [42, 43]. Another recent example is the use of gapmers to specifically degrade the mutant allele introduced by a mutation in the *COCH* gene, which causes autosomal dominant hearing impairment [44]. In this study, two strategies were used to degrade the mutant transcript: directly targeting the mutation or other single-nucleotide polymorphisms (SNPs) in *cis* with the mutation that are part of the mutant haplotype.

2.2.2 Disrupting Reading Frame

Splice-switching AONs can also be used to induce transcript degradation. Skipping regular exons can also be used to knockdown the function of an undesired gene, by creating mRNA isoforms that encode non-functional proteins or trigger degradation of the mRNA by NMD [45]. For instance, exon skipping of hepatic APOB100 was able to sustainably reduce LDL cholesterol levels in mice [46], downregulation of *MAPT* gene has been proposed as a possible treatment for tauopathies [47], and skipping exon 2 of *ALK5* may modulate the TGF-β signaling cascade, reducing the components related to the overproduction of extracellular matrix in hypertrophic scar [48].

3   Therapeutic Potential

While years ago oligonucleotides were considered a useful research tool and just a curiosity in the clinical market, this has rapidly changed into approved therapeutic strategies for several diseases and promising personalized treatments for many other (rare inherited) diseases. In this section, we will focus on the development of AON-based therapeutic strategies for two particular tissues: muscle and retina.

3.1   Examples of Clinical Trials for Muscle Diseases

The use of AONs to treat neuromuscular disorders has been at the forefront of the clinical development of AON-based therapies and more than half of the AONs currently in the market target either Duchenne muscular dystrophy (DMD) or spinal muscular atrophy (SMA). As previously described, AONs targeting the *DMD* gene aim to skip specific exons to restore the reading frame. This gene has 79 exons and patients present a large variety of mutations, mostly deletions and duplications, that require the design of specific AONs to treat a small subset to patients. The first such drug, eteplirsen, targeted exon 51 of *DMD*. Skipping this exon could potentially be therapeutic for 13% of DMD patients [3, 6]. Since then, golodirsen, viltolarsen, and casimersen have been approved, all applicable to decreasing percentages of patients [9, 49, 50].

All DMD exon-skipping drugs currently in the market are phosphorodiamidate morpholino oligomers (PMO). In contrast, the development of the first AON drug in clinical trials for this disorder, drisapersen (a 2'OMe/PS oligonucleotide) [51] as well as that of many others targeting *DMD* with the same chemistry were halted due to side effects [52]. Despite the apparent success of PMO chemistries to reach the market, these drugs are yet not very efficient, and their clinical outcomes are still poor. This is the main reason why they are yet to be approved in Europe, while in the USA and Japan have been given "accelerated approval" based on dystrophin protein expression as a surrogate endpoint, which is very low and there is debate about its clinical relevance [53]. Currently, several efforts are driven toward increasing the delivery efficacy of these drugs to the target tissue [32, 54]. Several next generation AONs targeting the same exon as eteplirsen (exon 51) have been or are being developed. This is the case of the stereopure suvodirsen, which was halted after poor results in a phase I clinical trial (NCT03907072) or the peptide-conjugated PMO currently in Phase I/II clinical trials (MOMENTUM, NCT04004065).

While AONs for DMD do not offer yet the clinical benefits that were hoped to achieve at initial stages, the journey to their development has provided very valuable lessons to stakeholders interested in developing these drugs, particularly in the context of orphan drugs [55]. A drug that benefited from some of the previous knowledge was nusinersen, a 2′MOE/PS AON targeting another neuromuscular disorder (SMA). Nusinersen was approved only months after eteplirsen and has been quickly approved worldwide due to the robust clinical data derived from the clinical trials [18, 56]. As described before, this AON is based on an exon inclusion approach to restore the expression of SMN protein in motoneurons. In this case, the target tissue is treated directly by intrathecal infusion, circumventing any delivery hurdles that may have hampered the efficacy of AONs targeting muscle or other organs when delivered systemically. Indeed, nusinersen's delivery approach, chosen chemistry and formulation has been replicated in several *n*-of-1 clinical trials of other AONs targeting motoneurons, such as milasen and jacifusen (NCT04768972) [57] (*see* Subheading 4).

3.2 Examples of Clinical Trials for Eye Diseases

The eye is one of the most promising organs for therapeutic development. Among other characteristics, it is contained, easily accessible, and immune-privileged [58]. In fact, the first-ever FDA-approved AON (fomiversen) was a first-class oligonucleotide to treat human cytomegalovirus retinitis, an eye condition in immunocompromised patients [33–36]. Furthermore, a growing group of genes and mutations causing retinal diseases have been targeted at preclinical level using AONs. This includes pseudoexon exclusion for *CEP290* [59–63], *OPA1* [64], *CHM* [65] *USH2A* [66], and *ABCA4* [24–29]; splicing modulation for *USH2A* [10] and *CEP290* [12]; or transcript degradation for *NR2E3* [67] and *RHO* [68]. Three of these molecules are currently in different clinical trial phases detailed below.

The most advanced molecule in a clinical setting is sepofarsen (QR-110). This is a 17-mer 2′OMe/PS oligonucleotide aiming to correct the inclusion of a pseudoexon caused by a deep-intronic mutation in *CEP290*-associated autosomal recessive Leber congenital amaurosis [69]. In the phase 1/2 clinical trial (NCT03140969), all patients were injected with an initial loading dose of either 320 or 160 µg followed by a maintenance dose every 3 months (160 or 80 µg) [70, 71]. Interim results showed that sepofarsen was well tolerated and safe with no serious adverse events [70, 71]. Although the final results of the trial have not yet been published, the improvement that most patients showed led to the design and approval of a phase 2/3 clinical trial (Illuminate,

NCT03913143). This is a multi-center, double-masked, randomized, controlled, multiple-dose study to evaluate efficacy, safety,
tolerability, and systemic exposure in patients older than 8 years
carrying the specific mutation in at least one of the two alleles. Two
different doses and a sham-procedure group will be assessed, for a
total period of 2 years. In addition, two other clinical trials for the
same molecule are ongoing. One is the extension of the phase 1/2
clinical trial to continue treating the patients of the first trial by
administering sepofarsen every 3 months in both the already inter-
vened and the contralateral eye (NCT03913130). The second is a
multi-center, open-label, dose-escalation, and double-masked randomized controlled trial to evaluate safety and tolerability in children below age of 8 years old (Brighten, NCT04855045).

A multi-center phase 1/2 clinical trial to assess safety and
tolerability of QR-421a (Stellar, NCT03780257) is currently
ongoing. This 21-mer 2′MOE/PS oligonucleotide aims to skip
the frequently mutated exon 13 of USH2A [10] causing autosomal
recessive Usher syndrome or isolated retinitis pigmentosa. Preliminary results, presented in a press release seem to indicate that
QR-421a is well tolerated with no serious adverse events. Furthermore, after treatment with this molecule, improvements in several
measures of vision were detected. With these encouraging results,
two preliminary phase 2/3 clinical trials have been designed in
order to study different patient populations based on the best
corrected visual acuity. Both trials will be double-masked, randomized, controlled, 24-month, and multiple-dose study (Sirius and
Celeste).

The third molecule in a clinical setting is QR-1123, a gapmer
designed to degrade the mutant allele (known as P23H) in the
RHO gene [68], which has a dominant-negative effect leading to
autosomal dominant retinitis pigmentosa. Thus, the hypothesis is
that by degrading the allele carrying the mutation, the other allele
will be able to produce a functional protein. This molecule is in an
early stage of a multi-center open-label, double-masked, randomized, phase 1/2 trial (NCT04123626).

Other molecules for eye-related genetic diseases in late stages of
preclinical development are QR-504a for TCF4-associated Fuchs
endothelial corneal dystrophy and QR-411 for pseudoexon exclusion in USH2A-associated Usher syndrome or isolated retinitis
pigmentosa.

As well as to target specific mutations, AONs have also been
explored for multifactorial eye conditions. This is the case of primary open angle glaucoma, in which TGF-β2 was targeted with a
14-mer 3 + 3 LNA-modified gapmer in a phase I clinical trial.
Results showed that the molecule was tolerated, safe and potentially
clinically efficacious [72]. Besides this, other type of antisense
molecules (small interference RNA, siRNA) have been clinically
tested for glaucoma [73], dry eye syndrome [74], diabetic macular
edema [75], and age-related macular degeneration [73, 76, 77].

4 Future of AON Trials and Personalized Medicine: $n = 1$ Trials?

In 2019, an AON molecule (milasen) to treat a single patient pushed the bounds of personalized medicine and raised many regulatory and ethical questions never explored before for genetic treatments [57, 78].

Milasen was customized exclusively for Mila, a child suffering from a form of Batten disease (neuronal ceroid lipofuscinosis 7) caused by the insertion of an SVA (SINE–VNTR–Alu) retrotransposon, with a detrimental effect on splicing, in the intron 6 of the *MFSD8* gene [57]. Using a 22-mer 2′MOE/PS AON it was possible to redirect splicing avoiding the insertion of the SVA in the final mRNA transcript. Besides the exclusivity of this treatment, another extraordinary achievement was that it took only 13 months to go from the clinical diagnosis to the first dosing: Mila had a clinical diagnosis in mid-November of 2016, the genetic defect was identified in May 2017, approval to proceed was received in January 2018 and first patient dosing occur in the same month.

The AON delivery regime via intrathecal bolus injection was highly similar to the one of nusinersen, the AON used for SMA [17, 18, 56]. The treatment did not show any safety concerns and the frequency and duration of the seizures was reduced. Unfortunately, despite the treatment had some effect, Mila passed away early 2021. Nevertheless, this study is the hallmark of personalized medicine, and although not all diseases are amenable for this type of therapies, it has highlighted this as a possible approach and managed to re-evaluate the speed and type of safety studies and regulatory requirements. In a similar development, a drug was designed and provided to patient suffering from amyotrophic lateral sclerosis (ALS) with mutations *FUS* gene, following the same delivery route as nusinersen and milasen. Unfortunately, this patient, Jaci Hermstad, also died recently. However, the drug originally developed for this single patient, ION363 or jacifusen, is currently being tested in a phase III trial for patients with the same disease (NCT04768972). Thus, AON technology can be considered as a platform for individualized treatments which may, sometimes, be extended to other patients.

5 Hurdles

A drawback when compared to small molecule drugs is the relatively large size of AON molecules which limits their delivery into the cells where they exert their action. Therefore, their distribution is limited, their naked uptake is poor, and it is highly determined by the chemistry of their backbones [54]. Often, these AON molecules are not even able to reach their target organ. To circumvent

88

Antisense Technology: Methods of Delivery and RNA Studies

this, most of these AONs rely on their conjugation or formulation with different delivery systems to be able to reach and access their intracellular targets [32]. In addition, when delivered systemically, these molecules can barely reach the central nervous system due to the blood retina and brain barriers. However, as described before, local delivery of naked modified AONs to these specific organs have shown to be efficient and safe in several clinical trials [17, 18, 56, 70, 71, 79].

Another drawback is the high exposure of certain organs upon systemic delivery of AONs. For instance, after intravenous injection of AONs a significant proportion is taken by the liver and kidney. This limits the biodistribution to other tissues and derivate on toxic effects in these organs. However, many of the liver and kidney injuries were found when using high and not clinically relevant doses of AONs [32]. In that sense, novel delivery methods or conjugates are required to be able to target the organs of interest and bypass the high clearance by the liver and kidneys.

Finding proper models to assess the sequence-dependent efficacy and safety of AONs is still a pending issue. Their safety assessment is often performed in rodents, non-human primates, and human plasma. However, these studies only provide sequence- and chemical modification-specific effects. The generation of humanized models have provided very good results, however, generating a humanized animal model for every mutation to be targeted is not feasible nor ethical. It is also possible to generate almost any human cell from patient-derived cells reprogrammed to a pluripotent stage. While these models can provide good readouts at RNA, protein, or even functional levels the entire context will still be missing. Currently, significant efforts are being made in the generation of organ-on-chips. This technology allows the combination of multiple tissues or even organs to study the interaction between them and test therapeutic interventions [80, 81]. In addition, this technology enables other type of measurements that in the near future might be very valuable to perform drug screenings and evaluate the efficacy and safety of many molecules, including AONs [80–83].

Finally, clear guidelines and novel clinical trial designs are needed to explore the full therapeutic potential of AONs when investigated as treatments for rare diseases. The case of milasen has proven that this is possible and new of such trials are being planned.

6 Conclusions

The therapeutic potential of AONs has been, for many years, subject of speculation and theoretical discussion and, while these molecules were widely applied in a research laboratory setting, their clinical application was anecdotal and limited to rare diseases. However, this landscape has recently changed completely thanks to several factors. On one hand, many of such drugs have been approved, being splice-switching AON and siRNA drugs at the forefront of this wave. Secondly, several breakthroughs in the delivery formulation of these drugs have increased the uptake of AONs targeting the liver and this has open wide open the field to consider these as reliable treatment options for several disorders where the liver is the target tissue. Thirdly, much more attention has been given to antisense technology due to the *n*-of-1 case of milasen. Lastly, RNA-therapies have gained extraordinary popularity due to vaccines against SAR-CoV-2 based on mRNA technology, highlighting the development of drugs based on nucleic acids. All of this will contribute to make these drugs a main resource in the therapeutic toolbox of the twenty-first century.

Acknowledgments

V.A-G holds a Miguel Servet Fellowship from the ISCIII (grants CP12/03057 and CPII17/00004), part-funded by ERDF/FEDER. V.A-G also acknowledges funding from Ikerbasque (Basque Foundation for Science). A.G. group is financially supported by the Foundation Fighting Blindness (PPA-0517-0717-RAD), the Curing Retinal Blindness Foundation as well as the Landelijke Stichting voor Blinden en Slechtzienden and Stichting Oogfonds via Uitzicht 2019-17, together with Stichting Blindenhulp, Rotterdamse Stichting Blindenbelangen and Dowilwo. The funding organizations had no role in the design or conduct of this research and provided unrestricted grants. Both authors are members of the European COST Action DARTER (CA17103).

References

1. Arechavala-Gomeza V, Khoo B, Aartsma-Rus A (2014) Splicing modulation therapy in the treatment of genetic diseases. Appl Clin Genet 7:245–252. https://doi.org/10.2147/TACG.S71506

2. Lewandowska MA (2013) The missing puzzle piece: splicing mutations. Int J Clin Exp Pathol 6(12):2675–2682

3. Arechavala-Gomeza V, Graham IR, Popplewell LJ, Adams AM, Aartsma-Rus A, Kinali M, Morgan JE, van Deutekom JC, Wilton SD, Dickson G, Muntoni F (2007) Comparative analysis of antisense oligonucleotide sequences for targeted skipping of exon 51 during dystrophin pre-mRNA splicing in human muscle. Hum Gene Ther 18(9):798–810. https://doi.org/10.1089/hum.2006.061

4. Popplewell LJ, Adkin C, Arechavala-Gomeza V, Aartsma-Rus A, de Winter CL, Wilton SD, Morgan JE, Muntoni F, Graham IR, Dickson G (2010) Comparative analysis of antisense oligonucleotide sequences targeting exon 53 of the human DMD gene: implications for future clinical trials. Neuromuscul Disord 20(2):102–110. https://doi.org/10.1016/j.nmd.2009.10.013

5. Kinali M, Arechavala-Gomeza V, Feng L, Cirak S, Hunt D, Adkin C, Guglieri M, Ashton E, Abbs S, Nihoyannopoulos P, Garralda ME, Rutherford M, McCulley C,

Popplewell L, Graham IR, Dickson G, Wood MJ, Wells DJ, Wilton SD, Kole R, Straub V, Bushby K, Sewry C, Morgan JE, Muntoni F (2009) Local restoration of dystrophin expression with the morpholino oligomer AVI-4658 in Duchenne muscular dystrophy: a single-blind, placebo-controlled, dose-escalation, proof-of-concept study. Lancet Neurol 8(10): 918–928. https://doi.org/10.1016/S1474-4422(09)70211-X

6. Cirak S, Arechavala-Gomeza V, Guglieri M, Feng L, Torelli S, Anthony K, Abbs S, Garralda ME, Bourke J, Wells DJ, Dickson G, Wood MJ, Wilton SD, Straub V, Kole R, Shrewsbury SB, Sewry C, Morgan JE, Bushby K, Muntoni F (2011) Exon skipping and dystrophin restoration in patients with Duchenne muscular dystrophy after systemic phosphorodiamidate morpholino oligomer treatment: an open-label, phase 2, dose-escalation study. Lancet 378(9791):595–605. https://doi.org/10.1016/S0140-6736(11)60756-3

7. Anwar S, Yokota T (2020) Golodirsen for Duchenne muscular dystrophy. Drugs Today 56(8):491–504. https://doi.org/10.1358/dot.2020.56.8.3159186

8. Clemens PR, Rao VK, Connolly AM, Harper AD, Mah JK, Smith EC, McDonald CM, Zaidman CM, Morgenroth LP, Osaki H, Satou Y, Yamashita T, Hoffman EP, Investigators CD (2020) Safety, tolerability, and efficacy of viltolarsen in boys with Duchenne muscular dystrophy amenable to exon 53 skipping: a phase 2 randomized clinical trial. JAMA Neurol 77: 1. https://doi.org/10.1001/jamaneurol.2020.1264

9. Shirley M (2021) Casimersen: first approval. Drugs 81:875. https://doi.org/10.1007/s40265-021-01512-2

10. Dulla K, Slijkerman R, van Diepen HC, Albert S, Dona M, Beumer W, Turunen JJ, Chan HL, Schulkens IA, Vorthoren L, Besten CD, Buil L, Schmidt I, Miao J, Venselaar H, Zang J, Neuhauss SCF, Peters T, Broekman S, Pennings R, Kremer H, Platenburg G, Adamson P, de Vrieze E, van Wijk E (2021) Antisense oligonucleotide-based treatment of retinitis pigmentosa caused by USH2A exon 13 mutations. Mol Ther 29:2441. https://doi.org/10.1016/j.ymthe.2021.04.024

11. Pendse N, Lamas V, Maeder M, Pawlyk B, Gloskowski S, Pierce EA, Chen Z-Y, Liu Q (2020) Exon 13-skipped USH2A protein retains functional integrity in mice, suggesting an exo-skipping therapeutic approach to treat USH2A-associated disease. bioRxiv. https://doi.org/10.1101/2020.02.04.934240

12. Barny I, Perrault I, Michel C, Goudin N, Defoort-Dhellemmes S, Ghazi I, Kaplan J, Rozet JM, Gerard X (2019) AON-mediated exon skipping to bypass protein truncation in retinal dystrophies due to the recurrent CEP290 c.4723A>T mutation. Fact or fiction? Genes (Basel) 10(5):368. https://doi.org/10.3390/genes10050368

13. Goto M, Sawamura D, Nishie W, Sakai K, McMillan JR, Akiyama M, Shimizu H (2006) Targeted skipping of a single exon harboring a premature termination codon mutation: implications and potential for gene correction therapy for selective dystrophic epidermolysis bullosa patients. J Investig Dermatol 126(12): 2614–2620. https://doi.org/10.1038/sj.jid.5700435

14. Bremer J, Bornert O, Nystrom A, Gostynski A, Jonkman MF, Aartsma-Rus A, van den Akker PC, Pasmooij AM (2016) Antisense oligonucleotide-mediated exon skipping as a systemic therapeutic approach for recessive dystrophic epidermolysis bullosa. Mol Ther Nucl Acids 5(10):e379. https://doi.org/10.1038/mtna.2016.87

15. Ham KA, Aung-Htut MT, Fletcher S, Wilton SD (2020) Nonsequential splicing events alter antisense-mediated exon skipping outcome in COL7A1. Int J Mol Sci 21(20):7705. https://doi.org/10.3390/ijms21207705

16. Bornert O, Hogervorst M, Nauroy P, Bischof J, Swildens J, Athanasiou I, Tufa SF, Keene DR, Kiritsi D, Hainzl S, Murauer EM, Marinkovich MP, Platenburg G, Hausser I, Wally V, Ritsema T, Koller U, Haisma EM, Nystrom A (2021) QR-313, an antisense oligonucleotide, shows therapeutic efficacy for treatment of dominant and recessive dystrophic epidermolysis bullosa: a preclinical study. J Investig Dermatol 141(4):883–893.e886. https://doi.org/10.1016/j.jid.2020.08.018

17. Finkel RS, Chiriboga CA, Vajsar J, Day JW, Montes J, De Vivo DC, Yamashita M, Rigo F, Hung G, Schneider E, Norris DA, Xia S, Bennett CF, Bishop KM (2016) Treatment of infantile-onset spinal muscular atrophy with nusinersen: a phase 2, open-label, dose-escalation study. Lancet 388(10063):3017–3026. https://doi.org/10.1016/S0140-6736(16)31408-8

18. Finkel RS, Mercuri E, Darras BT, Connolly AM, Kuntz NL, Kirschner J, Chiriboga CA, Saito K, Servais L, Tizzano E, Topaloglu H, Tulinius M, Montes J, Glanzman AM, Bishop K, Zhong ZJ, Gheuens S, Bennett CF, Schneider E, Farwell W, De Vivo DC, Group ES (2017) Nusinersen versus sham control in infantile-onset spinal muscular atrophy. N Engl

J Med 377(18):1723–1732. https://doi.org/
10.1056/NEJMoa1702752

19. Lentz JJ, Jodelka FM, Hinrich AJ, McCaffrey
KE, Farris HE, Spalitta MJ, Bazan NG, Duelli
DM, Rigo F, Hastings ML (2013) Rescue of
hearing and vestibular function by antisense
oligonucleotides in a mouse model of human
deafness. Nat Med 19(3):345–350. https://
doi.org/10.1038/nm.3106

20. Tomkiewicz TZ, Suárez-Herrera N, Cremers
FPM, Collin RWJ, Garanto A (2021) Antisense
oligonucleotide-based rescue of aberrant splicing defects caused by 15 pathogenic variants in
ABCA4. Int J Mol Sci 22(9):4621

21. Lim KH, Han Z, Jeon HY, Kach J, Jing E,
Weyn-Vanhentenryck S, Downs M,
Corrionero A, Oh R, Scharner J, Venkatesh A,
Ji S, Liau G, Ticho B, Nash H, Aznarez I
(2020) Antisense oligonucleotide modulation
of non-productive alternative splicing upregu-
lates gene expression. Nat Commun 11(1):
3501. https://doi.org/10.1038/s41467-
020-17093-9

22. Jaganathan K, Kyriazopoulou
Panagiotopoulou S, McRae JF, Darbandi SF,
Knowles D, Li YI, Kosmicki JA, Arbelaez J,
Cui W, Schwartz GB, Chow ED,
Kanterakis E, Gao H, Kia A, Batzoglou S, San-
ders SJ, Farh KK (2019) Predicting splicing
from primary sequence with deep learning.
Cell 176(3):535–548.e524. https://doi.org/
10.1016/j.cell.2018.12.015

23. Sangermano R, Khan M, Cornelis SS,
Richelle V, Albert S, Garanto A, Elmelik D,
Qamar R, Lugtenberg D, van den Born LI,
Collin RWJ, Cremers FPM (2018) ABCA4
midigenes reveal the full splice spectrum of all
reported noncanonical splice site variants in
Stargardt disease. Genome Res 28(1):
100–110. https://doi.org/10.1101/gr.
226621.117

24. Albert S, Garanto A, Sangermano R, Khan M,
Bax NM, Hoyng CB, Zernant J, Lee W,
Allikmets R, Collin RWJ, Cremers FPM
(2018) Identification and rescue of splice
defects caused by two neighboring deep-
intronic ABCA4 mutations underlying Stargardt disease. Am J Hum Genet 102(4):
517–527. https://doi.org/10.1016/j.ajhg.
2018.02.008

25. Bauwens M, Garanto A, Sangermano R,
Naessens S, Weisschuh N, De Zaeytijd J,
Khan M, Sadler F, Balikova I, Van
Cauwenbergh C, Rosseel T, Bauwens J, De
Leeneer K, De Jaegere S, Van Laethem T, De
Vries M, Carss K, Arno G, Fakin A, Webster
AR, de Ravel de l'Argentiere TJL, Sznajer Y,
Vuylsteke M, Kohl S, Wissinger B, Cherry T,

Collin RWJ, Cremers FPM, Leroy BP, De
Baere E (2019) ABCA4-associated disease as a
model for missing heritability in autosomal
recessive disorders: novel noncoding splice,
cis-regulatory, structural, and recurrent hypomorphic variants. Genet Med 21(8):
1761–1771. https://doi.org/10.1038/
s41436-018-0420-y

26. Garanto A, Duijkers L, Tomkiewicz TZ, Collin
RWJ (2019) Antisense oligonucleotide screening to optimize the rescue of the splicing defect
caused by the recurrent deep-intronic ABCA4
variant c.4539+2001G>A in Stargardt disease.
Genes 10(6):425. https://doi.org/10.3390/
genes10060452

27. Sangermano R, Garanto A, Khan M, Runhart
EH, Bauwens M, Bax NM, van den Born LI,
Khan MI, Cornelis SS, Verheij J, Pott JR,
Thiadens A, Klaver CCW, Puech B,
Meunier I, Naessens S, Arno G, Fakin A,
Carss KJ, Raymond FL, Webster AR, Dhaenens
CM, Stohr H, Grassmann F, Weber BHF,
Hoyng CB, De Baere E, Albert S, Collin
RWJ, Cremers FPM (2019) Deep-intronic
ABCA4 variants explain missing heritability in
Stargardt disease and allow correction of splice
defects by antisense oligonucleotides. Genet
Med 21(8):1751–1760. https://doi.org/10.
1038/s41436-018-0414-9

28. Khan M, Arno G, Fakin A, Parfitt DA, Dhooge
PPA, Albert S, Bax NM, Duijkers L,
Niblock M, Hau KL, Bloch E, Schiff ER,
Piccolo D, Hogden MC, Hoyng CB, Webster
AR, Cremers FPM, Cheetham ME, Garanto A,
Collin RWJ (2020) Detailed phenotyping and
therapeutic strategies for intronic ABCA4 var-
iants in Stargardt disease. Mol Ther Nucl Acids
21:412–427. https://doi.org/10.1016/j.
omtn.2020.06.007

29. Tomkiewicz TZ, Suarez-Herrera N, Cremers
FPM, Collin RWJ, Garanto A (2021) Antisense
oligonucleotide-based rescue of aberrant splicing defects caused by 15 pathogenic variants in
ABCA4. Int J Mol Sci 22(9):4621. https://
doi.org/10.3390/ijms22094621

30. Cremers FPM, Lee W, Collin RWJ, Allikmets R
(2020) Clinical spectrum, genetic complexity
and therapeutic approaches for retinal disease
caused by ABCA4 mutations. Prog Retin Eye
Res 79:100861. https://doi.org/10.1016/j.
preteyeres.2020.100861

31. Liang XH, Sun H, Nichols JG, Crooke ST
(2017) RNase H1-dependent antisense oligonucleotides are robustly active in directing
RNA cleavage in both the cytoplasm and the
nucleus. Mol Ther 25(9):2075–2092. https://
doi.org/10.1016/j.ymthe.2017.06.002

32. Hammond SM, Aartsma-Rus A, Alves S, Borgos SE, Buijsen RAM, Collin RWJ, Covello G, Denti MA, Desviat LR, Echevarria L, Foged C, Gaina G, Garanto A, Goyenvalle AT, Guzowska M, Holodnuka I, Jones DR, Krause S, Lehto T, Montolio M, Van Roon-Mom W, Arechavala-Gomeza V (2021) Delivery of oligonucleotide-based therapeutics: challenges and opportunities. EMBO Mol Med 13(4):e13243. https://doi.org/10.15252/emmm.202013243

33. Crooke ST (1998) Vitravene--another piece in the mosaic. Antis Nucl Acid Drug Dev 8(4): vii–viii. https://doi.org/10.1089/oli.1.1998.8.vii

34. Mulamba GB, Hu A, Azad RF, Anderson KP, Coen DM (1998) Human cytomegalovirus mutant with sequence-dependent resistance to the phosphorothioate oligonucleotide fomivirsen (ISIS 2922). Antimicrob Agents Chemother 42(4):971–973. https://doi.org/10.1128/AAC.42.4.971

35. Roehr B (1998) Fomivirsen approved for CMV retinitis. J Int Assoc Phys AIDS Care 4(10): 14–16

36. Jabs DA, Griffiths PD (2002) Fomivirsen for the treatment of cytomegalovirus retinitis. Am J Ophthalmol 133(4):552–556

37. Graham MJ, Lee RG, Bell TA III, Fu W, Mullick AE, Alexander VJ, Singleton W, Viney N, Geary R, Su J, Baker BF, Burkey J, Crooke ST, Crooke RM (2013) Antisense oligonucleotide inhibition of apolipoprotein C-III reduces plasma triglycerides in rodents, nonhuman primates, and humans. Circ Res 112(11): 1479–1490. https://doi.org/10.1161/CIRCRESAHA.111.300367

38. Geary RS, Baker BF, Crooke ST (2015) Clinical and preclinical pharmacokinetics and pharmacodynamics of mipomersen (kynamro ((R))): a second-generation antisense oligonucleotide inhibitor of apolipoprotein B. Clin Pharmacokinet 54(2):133–146. https://doi.org/10.1007/s40262-014-0224-4

39. Digenio A, Dunbar RL, Alexander VJ, Hompesch M, Morrow L, Lee RG, Graham MJ, Hughes SG, Yu R, Singleton W, Baker BF, Bhanot S, Crooke RM (2016) Antisense-mediated lowering of plasma apolipoprotein C-III by volanesorsen improves dyslipidemia and insulin sensitivity in type 2 diabetes. Diabetes Care 39(8):1408–1415. https://doi.org/10.2337/dc16-0126

40. Pechlaner R, Tsimikas S, Yin X, Willeit P, Baig F, Santer P, Oberhollenzer F, Egger G, Witztum JL, Alexander VJ, Willeit J, Kiechl S, Mayr M (2017) Very-low-density lipoprotein-associated apolipoproteins predict cardiovascular events and are lowered by inhibition of APOC-III. J Am Coll Cardiol 69(7): 789–800. https://doi.org/10.1016/j.jacc.2016.11.065

41. Aslesh T, Yokota T (2020) Development of antisense oligonucleotide gapmers for the treatment of dyslipidemia and lipodystrophy. Methods Mol Biol 2176:69–85. https://doi.org/10.1007/978-1-0716-0771-8_5

42. Benson MD, Waddington-Cruz M, Berk JL, Polydefkis M, Dyck PJ, Wang AK, Plante-Bordeneuve V, Barroso FA, Merlini G, Obici L, Scheinberg M, Brannagan TH III, Litchy WJ, Whelan C, Drachman BM, Adams D, Heitner SB, Conceicao I, Schmidt HH, Vita G, Campistol JM, Gamez J, Gorevic PD, Gane E, Shah AM, Solomon SD, Monia BP, Hughes SG, Kwoh TJ, McEvoy BW, Jung SW, Baker BF, Ackermann EJ, Gertz MA, Coelho T (2018) Inotersen treatment for patients with hereditary transthyretin amyloidosis. N Engl J Med 379(1):22–31. https://doi.org/10.1056/NEJMoa1716793

43. Dyck PJB, Coelho T, Waddington Cruz M, Brannagan TH III, Khella S, Karam C, Berk JL, Polydefkis MJ, Kincaid JC, Wiesman JF, Litchy WJ, Mauermann ML, Ackermann EJ, Baker BF, Jung SW, Guthrie S, Pollock M, Dyck PJ (2020) Neuropathy symptom and change: inotersen treatment of hereditary transthyretin amyloidosis. Muscle Nerve 62(4):509–515. https://doi.org/10.1002/mus.27023

44. de Vrieze E, Canas Martin J, Peijnenborg J, Martens A, Oostrik J, van den Heuvel S, Neveling K, Pennings R, Kremer H, van Wijk E (2021) AON-based degradation of c.151C>T mutant COCH transcripts associated with dominantly inherited hearing impairment DFNA9. Mol Ther Nucl Acids 24:274–283. https://doi.org/10.1016/j.omtn.2021.02.033

45. Maquat LE (2004) Nonsense-mediated mRNA decay: splicing, translation and mRNP dynamics. Nat Rev Mol Cell Biol 5(2):89–99. https://doi.org/10.1038/nrm1310

46. Disterer P, Al-Shawi R, Ellmerich S, Waddington SN, Owen JS, Simons JP, Khoo B (2013) Exon skipping of hepatic APOB pre-mRNA with splice-switching oligonucleotides reduces LDL cholesterol in vivo. Mol Ther 21(3): 602–609. https://doi.org/10.1038/mt.2012.264

47. Sud R, Geller ET, Schellenberg GD (2014) Antisense-mediated exon skipping decreases tau protein expression: a potential therapy for tauopathies. Mol Ther Nucl Acids 3:e180. https://doi.org/10.1038/mtna.2014.30

48. Raktoe RS, Rietveld MH, Out-Luiting JJ, Kruithof-de Julio M, van Zuijlen PP, van Doorn R, Ghalbzouri AE (2020) Exon skipping of TGFbetaRI affects signalling and ECM expression in hypertrophic scar-derived fibroblasts. Scars Burn Heal 6: 2059513120908857. https://doi.org/10.1177/2059513120908857

49. Aartsma-Rus A, Corey DR (2020) The 10th oligonucleotide therapy approved: golodirsen for Duchenne muscular dystrophy. Nucl Acids Ther 30(2):67–70. https://doi.org/10.1089/nat.2020.0845

50. Dhillon S (2020) Viltolarsen: first approval. Drugs 80(10):1027–1031. https://doi.org/10.1007/s40265-020-01339-3

51. van Deutekom JC, Janson AA, Ginjaar IB, Frankhuizen WS, Aartsma-Rus A, Bremmer-Bout M, den Dunnen JT, Koop K, van der Kooi AJ, Goemans NM, de Kimpe SJ, Ekhart PF, Venneker EH, Platenburg GJ, Verschuuren JJ, van Ommen GJ (2007) Local dystrophin restoration with antisense oligonucleotide PRO051. N Engl J Med 357(26): 2677–2686. https://doi.org/10.1056/NEJMoa073108

52. Hilhorst N, Spanoudi-Kitrimi I, Goemans N, Morren MA (2019) Injection site reactions after long-term subcutaneous delivery of drisapersen: a retrospective study. Eur J Pediatr 178(2):253–258. https://doi.org/10.1007/s00431-018-3272-1

53. Aartsma-Rus A, Arechavala-Gomeza V (2018) Why dystrophin quantification is key in the eteplirsen saga. Nat Rev Neurol 14(8): 454–456. https://doi.org/10.1038/s41582-018-0033-8

54. Godfrey C, Desviat LR, Smedsrod B, Pietri-Rouxel F, Denti MA, Disterer P, Lorain S, Nogales-Gadea G, Sardone V, Anwar R, El Andaloussi S, Lehto T, Khoo B, Brolin C, van Roon-Mom WM, Goyenvalle A, Aartsma-Rus A, Arechavala-Gomeza V (2017) Delivery is key: lessons learnt from developing splice-switching antisense therapies. EMBO Mol Med 9(5):545–557. https://doi.org/10.15252/emmm.201607199

55. Straub V, Balabanov P, Bushby K, Ensini M, Goemans N, De Luca A, Pereda A, Hemmings R, Campion G, Kaye E, Arechavala-Gomeza V, Goyenvalle A, Niks E, Veldhuizen O, Furlong P, Stoyanova-Beninska-V, Wood MJ, Johnson A, Mercuri E, Muntoni F, Sepodes B, Haas M, Vroom E, Aartsma-Rus A (2016) Stakeholder cooperation to overcome challenges in orphan medicine development: the example of Duchenne muscular dystrophy. Lancet Neurol 15(8): 882–890. https://doi.org/10.1016/S1474-4422(16)30035-7

56. Mercuri E, Darras BT, Chiriboga CA, Day JW, Campbell C, Connolly AM, Iannaccone ST, Kirschner J, Kuntz NL, Saito K, Shieh PB, Tulinius M, Mazzone ES, Montes J, Bishop KM, Yang Q, Foster R, Gheuens S, Bennett CF, Farwell W, Schneider E, De Vivo DC, Finkel RS, Group CS (2018) Nusinersen versus sham control in later-onset spinal muscular atrophy. N Engl J Med 378(7):625–635. https://doi.org/10.1056/NEJMoa1710504

57. Kim J, Hu C, Moufawad El Achkar C, Black LE, Douville J, Larson A, Pendergast MK, Goldkind SF, Lee EA, Kuniholm A, Soucy A, Vaze J, Belur NR, Fredriksen K, Stojkovska I, Tsytsykova A, Armant M, DiDonato RL, Choi J, Cornelissen L, Pereira LM, Augustine EF, Genetti CA, Dies K, Barton B, Williams L, Goodlett BD, Riley BL, Pasternak A, Berry ER, Pflock KA, Chu S, Reed C, Tyndall K, Agrawal PB, Beggs AH, Grant PE, Urion DK, Snyder RO, Waisbren SE, Poduri A, Park PJ, Patterson A, Biffi A, Mazzulli JR, Bodamer O, Berde CB, Yu TW (2019) Patient-customized oligonucleotide therapy for a rare genetic disease. N Engl J Med 381(17):1644–1652. https://doi.org/10.1056/NEJMoa1813279

58. Vazquez-Dominguez I, Garanto A, Collin RWJ (2019) Molecular therapies for inherited retinal diseases-current standing, opportunities and challenges. Genes 10(9):654. https://doi.org/10.3390/genes10090654

59. Collin RW, den Hollander AI, van der Velde-Visser SD, Bennicelli J, Bennett J, Cremers FP (2012) Antisense oligonucleotide (AON)-based therapy for leber congenital amaurosis caused by a frequent mutation in CEP290. Mol Ther Nucl Acids 1:e14. https://doi.org/10.1038/mtna.2012.3

60. Gerard X, Perrault I, Hanein S, Silva E, Bigot K, Defoort-Delhemmes S, Rio M, Munnich A, Scherman D, Kaplan J, Kichler A, Rozet JM (2012) AON-mediated exon skipping restores ciliation in fibroblasts harboring the common leber congenital amaurosis CEP290 mutation. Mol Ther Nucl Acids 1: e29. https://doi.org/10.1038/mtna.2012.21

61. Garanto A, Chung DC, Duijkers L, Corral-Serrano JC, Messchaert M, Xiao R, Bennett J, Vandenberghe LH, Collin RW (2016) In vitro and in vivo rescue of aberrant splicing in CEP290-associated LCA by antisense oligonucleotide delivery. Hum Mol Genet 25(12): 2552–2563. https://doi.org/10.1093/hmg/ddw118

62. Duijkers L, van den Born LI, Neidhardt J, Bax NM, Pierrache LHM, Klevering BJ, Collin

RWJ, Garanto A (2018) Antisense oligonucleotide-based splicing correction in individuals with leber congenital amaurosis due to compound heterozygosity for the c.2991+1655A>G mutation in CEP290. Int J Mol Sci 19(3):753. https://doi.org/10.3390/ijms19030753

63. Dulla K, Aguila M, Lane A, Jovanovic K, Parfitt DA, Schulkens I, Chan HL, Schmidt I, Beumer W, Vorthoren L, Collin RWJ, Garanto A, Duijkers L, Brugulat-Panes A, Semo M, Vugler AA, Biasutto P, Adamson P, Cheetham ME (2018) Splice-modulating oligonucleotide QR-110 restores CEP290 mRNA and function in human c.2991+1655A>G LCA10 models. Mol Ther Nucl Acids 12:730–740. https://doi.org/10.1016/j.omtn.2018.07.010

64. Bonifert T, Gonzalez Menendez I, Battke F, Theurer Y, Synofzik M, Schols L, Wissinger B (2016) Antisense oligonucleotide mediated splice correction of a deep intronic mutation in OPA1. Mol Ther Nucl Acids 5(11):e390. https://doi.org/10.1038/mtna.2016.93

65. Garanto A, van der Velde-Visser SD, Cremers FPM, Collin RWJ (2018) Antisense oligonucleotide-based splice correction of a deep-intronic mutation in CHM underlying choroideremia. Adv Exp Med Biol 1074:83–89. https://doi.org/10.1007/978-3-319-75402-4_11

66. Slijkerman RW, Vache C, Dona M, Garcia-Garcia G, Claustres M, Hetterschijt L, Peters TA, Hartel BP, Pennings RJ, Millan JM, Aller E, Garanto A, Collin RW, Kremer H, Roux AF, Van Wijk E (2016) Antisense oligonucleotide-based splice correction for USH2A-associated retinal degeneration caused by a frequent deep-intronic mutation. Mol Ther Nucl Acids 5(10):e381. https://doi.org/10.1038/mtna.2016.89

67. Naessens S, Ruysschaert L, Lefever S, Coppieters F, De Baere E (2019) Antisense oligonucleotide-based downregulation of the G56R pathogenic variant causing NR2E3-associated autosomal dominant retinitis pigmentosa. Genes 10(5):363. https://doi.org/10.3390/genes10050363

68. Murray SF, Jazayeri A, Matthes MT, Yasumura D, Yang H, Peralta R, Watt A, Freier S, Hung G, Adamson PS, Guo S, Monia BP, LaVail MM, McCaleb ML (2015) Allele-specific inhibition of rhodopsin with an antisense oligonucleotide slows photoreceptor cell degeneration. Invest Ophthalmol Vis Sci 56(11):6362–6375. https://doi.org/10.1167/iovs.15-16400

69. den Hollander AI, Koenekoop RK, Yzer S, Lopez I, Arends ML, Voesenek KE, Zonneveld MN, Strom TM, Meitinger T, Brunner HG, Hoyng CB, van den Born LI, Rohrschneider K, Cremers FPM (2006) Mutations in the CEP290 (NPHP6) gene are a frequent cause of Leber congenital amaurosis. Am J Hum Genet 79(3):556–561. https://doi.org/10.1086/507318

70. Cideciyan AV, Jacobson SG, Drack AV, Ho AC, Charng J, Garafalo AV, Roman AJ, Sumaroka A, Han IC, Hochstedler MD, Pfeifer WL, Sohn EH, Taiel M, Schwartz MR, Biasutto P, Wit W, Cheetham ME, Adamson P, Rodman DM, Platenburg G, Tome MD, Balikova I, Nerinckx F, Zaeytijd J, Van Cauwenbergh C, Leroy BP, Russell SR (2019) Effect of an intravitreal antisense oligonucleotide on vision in Leber congenital amaurosis due to a photoreceptor cilium defect. Nat Med 25(2):225–228. https://doi.org/10.1038/s41591-018-0295-0

71. Cideciyan AV, Jacobson SG, Ho AC, Garafalo AV, Roman AJ, Sumaroka A, Krishnan AK, Swider M, Schwartz MR, Girach A (2021) Durable vision improvement after a single treatment with antisense oligonucleotide sepofarsen: a case report. Nat Med 27:785. https://doi.org/10.1038/s41591-021-01297-7

72. Pfeiffer N, Voykov B, Renieri G, Bell K, Richter P, Weigel M, Thieme H, Wilhelm B, Lorenz K, Feindor M, Wosikowski K, Janicot M, Packert D, Rommich R, Mala C, Fettes P, Leo E (2017) First-in-human phase I study of ISTH0036, an antisense oligonucleotide selectively targeting transforming growth factor beta 2 (TGF-beta2), in subjects with open-angle glaucoma undergoing glaucoma filtration surgery. PLoS One 12(11):e0188899. https://doi.org/10.1371/journal.pone.0188899

73. Moreno-Montanes J, Sadaba B, Ruz V, Gomez-Guiu A, Zarranz J, Gonzalez MV, Paneda C, Jimenez AI (2014) Phase I clinical trial of SYL040012, a small interfering RNA targeting beta-adrenergic receptor 2, for lowering intraocular pressure. Mol Ther 22(1):226–232. https://doi.org/10.1038/mt.2013.217

74. Benitez-Del-Castillo JM, Moreno-Montanes J, Jimenez-Alfaro I, Munoz-Negrete FJ, Turman K, Palumaa K, Sadaba B, Gonzalez MV, Ruz V, Vargas B, Paneda C, Martinez T, Bleau AM, Jimenez AI (2016) Safety and efficacy clinical trials for SYL1001, a novel short interfering RNA for the treatment of dry eye disease. Invest Ophthalmol Vis Sci 57(14):

6447–6454. https://doi.org/10.1167/iovs.16-20303

75. Nguyen QD, Schachar RA, Nduaka CI, Sperling M, Basile AS, Klamerus KJ, Chi-Burris K, Yan E, Paggiarino DA, Rosenblatt I, Aitchison R, Erlich SS, Group DCS (2012) Dose-ranging evaluation of intravitreal siRNA PF-04523655 for diabetic macular edema (the DEGAS study). Invest Ophthalmol Vis Sci 53(12):7666–7674. https://doi.org/10.1167/iovs.12-9961

76. Kaiser PK, Symons RC, Shah SM, Quinlan EJ, Tabandeh H, Do DV, Reisen G, Lockridge JA, Short B, Guerciolini R, Nguyen QD, Sirna-027 Study I (2010) RNAi-based treatment for neovascular age-related macular degeneration by Sirna-027. Am J Ophthalmol 150(1):33–39.e32. https://doi.org/10.1016/j.ajo.2010.02.006

77. Nguyen QD, Schachar RA, Nduaka CI, Sperling M, Klamerus KJ, Chi-Burris K, Yan E, Paggiarino DA, Rosenblatt I, Aitchison R, Erlich SS, Group MCS (2012) Evaluation of the siRNA PF-04523655 versus ranibizumab for the treatment of neovascular age-related macular degeneration (MONET Study). Ophthalmology 119(9):1867–1873. https://doi.org/10.1016/j.ophtha.2012.03.043

78. Mullard A (2020) N-of-1 drugs push biopharma frontiers. Nat Rev Drug Discov 19(3):151–153. https://doi.org/10.1038/d41573-020-00027-x

79. Tabrizi SJ, Leavitt BR, Landwehrmeyer GB, Wild EJ, Saft C, Barker RA, Blair NF, Craufurd D, Priller J, Rickards H, Rosser A, Kordasiewicz HB, Czech C, Swayze EE, Norris DA, Baumann T, Gerlach I, Schobel SA, Paz E, Smith AV, Bennett CF, Lane RM, Phase 1-2a I-HSST (2019) Targeting huntingtin expression in patients with Huntington's disease. N Engl J Med 380(24):2307–2316. https://doi.org/10.1056/NEJMoa1900907

80. Bhise NS, Ribas J, Manoharan V, Zhang YS, Polini A, Massa S, Dokmeci MR, Khademhosseini A (2014) Organ-on-a-chip platforms for studying drug delivery systems. J Control Release 190:82–93. https://doi.org/10.1016/j.jconrel.2014.05.004

81. Kimura H, Sakai Y, Fujii T (2018) Organ/body-on-a-chip based on microfluidic technology for drug discovery. Drug Metab Pharmacokinet 33(1):43–48. https://doi.org/10.1016/j.dmpk.2017.11.003

82. Mittal R, Woo FW, Castro CS, Cohen MA, Karanxha J, Mittal J, Chhibber T, Jhaveri VM (2019) Organ-on-chip models: implications in drug discovery and clinical applications. J Cell Physiol 234(6):8352–8380. https://doi.org/10.1002/jcp.27729

83. Jodat YA, Kang MG, Kiaee K, Kim GJ, Martinez AFH, Rosenkranz A, Bae H, Shin SR (2018) Human-derived organ-on-a-chip for personalized drug development. Curr Pharm Des 24(45):5471–5486. https://doi.org/10.2174/1381612825666190308150055

7

In Vitro Models for the Evaluation of Antisense Oligonucleotides in Skin

Jeroen Bremer and Peter C. van den Akker

Abstract

The genodermatosis dystrophic epidermolysis bullosa (DEB) is caused by mutations in the *COL7A1* gene which encodes type VII collagen (C7). In the cutaneous basement membrane zone, C7 secures attachment of the epidermal basal keratinocyte to the papillary dermis by means of anchoring fibril formation. The complete absence of these anchoring fibrils leads to severe blistering of skin and mucosa upon the slightest friction and early mortality. To date, although preclinical advances toward therapy are promising, treatment for the disease is merely symptomatic. Therefore, research into novel therapeutics is warranted.

Antisense oligonucleotide (ASO)-mediated exon skipping is such a therapy. Clinical examination of naturally occurring exon skipping suggested that this mechanism could most likely benefit the most severely affected patients. The severe form of DEB is caused by biallelic null mutations. Exon skipping aims to bind an ASO to the mutated exon of the pre-mRNA in the cell nucleus. Thereby, the ASO inhibits the recognition of the mutated exon by the splicing machinery, and as a result, the mutated exon is spliced out from the mRNA with its surrounding introns, i.e., it is skipped. Here, we describe in vitro methods to evaluate ASO-mediated exon skipping in a preclinical setting.

Key words Epidermolysis bullosa, Therapy, Exon skipping, Antisense RNA, Fibroblasts, Keratinocytes, Splice modulating

1 Introduction

In this chapter, we describe the evaluation of ASO-mediated exon skipping as a therapeutic approach for DEB in a preclinical in vitro setting. DEB is caused by mutations in the *COL7A1* gene which encodes type VII collagen (C7) [1]. DEB is a rare disease affecting 1–9 in every one million births, worldwide. The disease is characterized by severe blistering of skin and mucosae. DEB can be inherited both dominantly and recessively, and the severity of the disease strongly depends on the quantity and functionality of the C7 protein present at the cutaneous basement membrane zone. The most severe recessive form of DEB (RDEB-gen sev) is caused by biallelic null mutations and the complete absence of C7 in the skin. Previously, we have shown that for RDEB-gen sev caused by biallelic null mutations, exon skipping is anticipated to be clinically beneficial [2].

Exon skipping relies on specifically designed ASOs that bind to the pre-mRNA in the cell nucleus. When bound, these ASOs inhibit the recognition of the mutated exon by the splicing machinery through steric hindrance [3]. As a result, the mutated exon is spliced out (skipped) of the mRNA together with its surrounding introns. If the skipped exon is in frame, the reading frame of the transcript is maintained and produces a slightly shorter but functional protein [4].

Exon skipping affects the pre-mRNA; therefore, it is essential for the ASO to pass the cell membrane and the nuclear envelope. The commonly used 2′-O-methyl phosphorothioate (2OMePS) and 2′-methoxyethyl phosphorothioate (2MOE) ASOs are negatively charged and therefore not able to easily pass the cell membrane in cell cultures. Therefore, active transfer across the cell membrane is essential. Cationic lipid transfection is such a way of active transfer and widely used to achieve efficient uptake by in vitro cultured cells. Widely studied cells of the skin are dermal fibroblasts and epidermal keratinocytes. Here, we describe the in vitro evaluation of distribution and activity of ASOs in cultured fibroblasts and keratinocytes, as C7 is expressed by both the cell types. However, cationic lipid transfection of fibroblasts and keratinocytes can be used to evaluate the activity of antisense RNA for many diseases, as they express many proteins.

2 Materials

2.1 Cell Culture

1. Trypsin/EDTA (2.5% trypsin/0.2% EDTA).
2. Dispase II (2.4 U/mL).
3. Penicillin/Streptomycin (100 U/mL and 100 μg/mL, respectively).
4. Saline solution: 0.9% NaCl in dH$_2$O sterilized through 0.22-μm filter.
5. Antisense oligonucleotides: 50 μM stock solution in dH$_2$O (final concentration depends on experimental setup).
6. Fibroblast medium: DMEM 4.5 g/L glucose, L-glutamate, 10% fetal bovine serum (FBS), 1× penicillin/streptomycin.
7. Phosphate-buffered saline (PBS).
8. Fibroblast transfection agent: Polyethylenimine (PEI) 1 mg/mL.
9. Keratinocyte medium: CellnTec Prime (CnT-PR).
10. HEPES-buffered saline solution (HBSS).

11. Keratinocyte transfection medium: Opti-MEM.

12. Keratinocyte transfection agent: Lipofectamine-2000.

3 Methods

3.1 Isolation and Culture of Epidermal Keratinocytes and Dermal Fibroblasts

Full-thickness skin biopsies (4–6 mm) or larger skin tissue (1–2 cm) are used to isolate cells.

1. On day 1, incubate the tissue at room temperature overnight in 2× penicillin/streptomycin solution protected from light.

2. On day 2, using tweezers and a scalpel, scrape off excess fatty tissue from the dermal side of the tissue.

3. Place the tissue in a 100-mm petri dish, floating with the dermal side down in 10 mL Dispase II and incubated overnight at 4 °C.

4. On day 3, separate the epidermis from the dermis as a sheet using tweezers. After separation, place the dermis into a glass petri dish and set aside. Transfer the epidermal sheets into a clean 100-mm petri dish containing 10 mL trypsin/EDTA and incubate for 10 min at 37 °C.

5. During the 10 min incubation time, cut the dermis into small pieces of around 1–2 mm in size, using two scalpels in a "scissor" fashion on the glass surface of the petri dishes. Transfer the tissue fragments onto the bottom of the well of a six-well plate and add complete fibroblast medium dropwise onto the tissue and refresh every other day. Within 2 weeks, fibroblasts should be growing out from the tissue. When proliferation of these fibroblasts starts to stagnate due to confluency, remove the tissue remnants, and harvest and passage the cells.

6. After 10 min trypsinization of the epidermal sheets, pipette the trypsin/EDTA solution up and down repeatedly to dissociate the sheets into individual cells.

7. Transfer the cell suspension into a 15-mL tube containing 500 μL FBS to inactivate trypsinization.

8. Spin the cells down by centrifugation for 10 min at 200 × g.

9. Discard the supernatant and resuspend the pellet in Cnt-PR medium.

10. Seed the cells into the culture vessel. Usually, keratinocytes isolated from one 6 mm biopsy are seeded in two 35-mm petri dishes. Refresh the medium three times a week and harvest and passage the cells once they reach 75–90% confluence.

3.2 Transfection of Primary Fibroblasts and Keratinocytes

In this protocol, we describe the transfection of cells with an ASO in a well of a 12-well plate at a concentration of 250 nM, as an example. This is the final concentration of ASOs in the wells after the transfection. All concentrations can be adjusted according to the needs of the individual experiments.

3.2.1 Fibroblasts

Fibroblasts are cultured and transfected in normal fibroblast medium (*see* Subheading 2.1, **step 6**)

1. Seed the fibroblasts in 12-well plates at a density at which the cells reach 70–80% confluence within 24–48 h (*see* **Note 1**), depending on proliferation rate.

2. When the cells reach 70–80% confluence, wash the cells three times with PBS and carefully remove the PBS with a pipet or vacuum aspiration system.

3. Add 900 μL fresh medium and place the plate back in the incubator (*see* **Note 2**).

4. To prepare the lipid-ASO complexes, for each transfected well, pipette 91.5 μL sterile saline solution in a sterile Eppendorf cup.

5. Add 5 μL of ASO, and 3.5 μL PEI and immediately vortex for 10 s (*see* **Note 3**).

6. Incubate the saline-ASO-PEI solution at room temperature for 10 min.

7. After incubation, add the transfection mix dropwise to the well.

8. Shake the plate in a north to south and east to west motion and place in the incubator (*see* **Note 4**).

9. After 5–6 h, gently remove the transfection medium, wash twice with PBS, and add 1 mL fresh medium and place back into the incubator (*see* **Note 5**).

10. After 48–72 h, analyze the cells for exon skipping at RNA or protein level.

11. As a positive control, a fluorescently labeled nonspecific AON is used. In case exon skipping exerts its effect in the nuclei, localization in nuclei corresponds to transfection efficiency (Fig. 1).

3.2.2 Keratinocytes

Keratinocytes are cultured in CnT-PR serum-free low-calcium keratinocyte medium and transfected in Opti-MEM medium.

1. Seed keratinocytes in 12-well plates at a density at which the cells reach 70–80% confluence within 24–48 h.

2. When the cells reach 70–80% confluence, wash the cells three times with HBSS and add 900 μL Opti-MEM to the well and place back in the incubator.

Fibroblasts   Keratinocytes

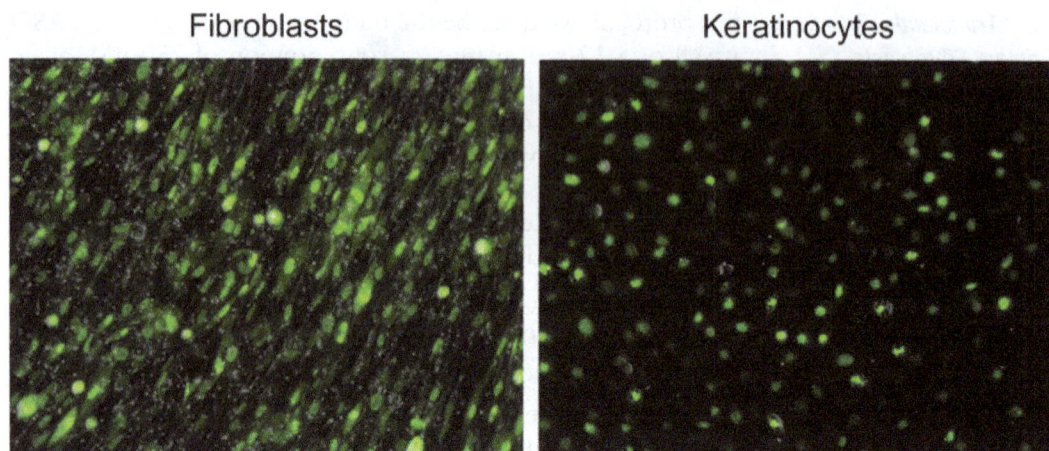

Fig. 1 Microscopy image of cells transfected with fluorescently labeled AON. **Left**: Fibroblasts transfected with a fluorescently labeled (green) AON using polyethylenimine. **Right**: Keratinocytes transfected with the same fluorescently labeled AON. A transfection efficiency of more than 95% is observed in both fibroblasts and keratinocytes, as shown by the green signal in the nuclei

3. In an 1.5-mL tube, add 45 μL Opti-MEM, then add 5 μL ASO and gently mix by pipetting in and out.

4. In a second 1.5-mL tube, add 48 μL Opti-MEM and 2 μL Lipofectamine-2000 and gently mix by pipetting in and out and incubate at room temperature for 5 min.

5. Add the Lipofectamin-2000 solution to the ASO solution and gently mix by pipetting up and down.

6. Incubate at room temperature for 30 min.

7. Add the lipid-ASO complexes dropwise to the wells and place back in the incubator.

8. After 6 h of incubation, remove the medium and gently wash the cells twice with HBSS. Add fresh keratinocyte medium and place the plate in the incubator.

9. After 24–72 h cells can be analyzed for exon skipping and protein expression.

4   Notes

1. For both fibroblasts and keratinocytes: usually between 0.5 and 1.5×10^5 of primary cultured cells is sufficient depending on passage and viability.

2. Do not pipet the PBS, HBSS, or medium directly onto the cells. Instead, gently pipet the liquid against the side wall of the well to prevent unnecessary stress to the cells.

3. It is essential to pipet in the order saline, ASO, PEI. Do not vortex longer than 10 s.

4. During incubation, prevent the solution from agitating. When pipetting the solution from the Eppendorf tube onto the wells, do not pipet repeatedly up and down, as this might disrupt the lipid-ASO complexes. Additionally, when placing the plate back into the incubator, prevent swirling motions, as this will concentrate the AON-transfection reagent complexes in the middle of the wells.

5. Transfection using cationic lipids induces, to some extent, cell death and will have an effect on the cell membrane as the lipids bind and pass them. Therefore, when washing cells and refreshing media, gently and smoothly pipet via the side of the well.

References

1. Has C, Bauer JW, Bodemer C, Bolling MC, Bruckner-Tuderman L, Diem A, Fine JD, Heagerty A, Hovnanian A, Marinkovich MP, Martinez AE, McGrath JA, Moss C, Murrell DF, Palisson F, Schwieger-Briel A, Sprecher E, Tamai K, Uitto J, Woodley DT, Zambruno G, Mellerio JE (2020) Consensus reclassification of inherited epidermolysis bullosa and other disorders with skin fragility. Br J Dermatol 183(4): 614–627. https://doi.org/10.1111/bjd.18921

2. Bremer J, van der Heijden EH, Eichhorn DS, Meijer R, Lemmink HH, Scheffer H, Sinke RJ, Jonkman MF, Pasmooij AMG, Van den Akker PC (2019) Natural exon skipping sets the stage for exon skipping as therapy for dystrophic epidermolysis bullosa. Mol Ther Nucleic Acids 18: 465–475

3. Aartsma-Rus A, van Vliet L, Hirschi M, Janson AA, Heemskerk H, de Winter CL, de Kimpe S, van Deutekom JC, t Hoen PA, van Ommen GJ (2009) Guidelines for antisense oligonucleotide design and insight into splice-modulating mechanisms. Mol Ther 17(3):548–553. https://doi.org/10.1038/mt.2008.205

4. Bornert O, Kühl T, Bremer J, van den Akker PC, Pasmooij AM, Nyström A (2016) Analysis of the functional consequences of targeted exon deletion in COL7A1 reveals prospects for dystrophic epidermolysis bullosa therapy. Mol Ther 24(7): 1302–1311. https://doi.org/10.1038/mt.2016.92

8

How to Design U1 snRNA Molecules for Splicing Rescue

Liliana Matos, Juliana I. Santos, Mª. Francisca Coutinho
and Sandra Alves

Abstract

Mutations affecting constitutive splice donor sites (5′ss) are among the most frequent genetic defects that disrupt the normal splicing process. Pre-mRNA splicing requires the correct identification of a number of *cis*-acting elements in an ordered fashion. By disrupting the complementarity of the 5′ss with the endogenous small nuclear RNA U1 (U1 snRNA), the key component of the spliceosomal U1 ribonucleoprotein, 5′ss mutations may result in exon skipping, intron retention or activation of cryptic splice sites. Engineered modification of the U1 snRNA seemed to be a logical method to overcome the effect of those mutations. In fact, over the last years, a number of in vitro studies on the use of those modified U1 snRNAs to correct a variety of splicing defects have demonstrated the feasibility of this approach. Furthermore, recent reports on its applicability in vivo are adding up to the principle that engineered modification of U1 snRNAs represents a valuable approach and prompting further studies to demonstrate the clinical translatability of this strategy.

Here, we outline the design and generation of U1 snRNAs with different degrees of complementarity to mutated 5′ss. Using the *HGSNAT* gene as an example, we describe the methods for a proper evaluation of their efficacy in vitro, taking advantage of our experience to share a number of tips on how to design U1 snRNA molecules for splicing rescue.

Key words U1 snRNA-based therapy, Splicing modulation, 5′ss mutations, Aberrant exon skipping, Modified U1 snRNA, Mucopolysaccharidosis IIIC

---

1 Introduction

The U1 small nuclear ribonucleoprotein (U1 snRNP) is a key molecule involved in an early event of the splicing process. Like other snRNPs involved in the overall splicing regulation process, it contains a small RNA complexed with several proteins, namely seven Smith antigen (Sm) proteins and three U1-specific proteins (U1A, U1C, and U170K) [1]. U1 snRNA, the RNA component of the U1 snRNP is a 164 nucleotides-long molecule whose 5′ end interacts by complementarity with the 5′ splice donor site (5′ss). That interaction between the single stranded 5′ tail of the U1 snRNA molecule and the moderately conserved stretch of nucleotides that constitutes the 5′ss (CAG/GURAGU, where R is a purine) marks the exon-intron boundary and initiates spliceosome assembly [2]. About 40%, 22%, and 5% of normal 5′ss contain two, three, or four mismatches towards the U1 snRNA, respectively [3, 4]. This variable degree of degeneration is among the major factors that significantly contribute to hinder a clear prediction of the effect of mutations flanking the canonical GU site. Furthermore, there is a number of additional elements, which may influence the splice site selection and need to be taken into account such as splicing silencer and enhancer motifs, the presence of alternative splice sites, secondary structures, and regulatory proteins [5]. Therefore, a straightforward prediction of the effect of mutations flanking the canonical GU site without a direct assessment of the mature mRNA produced can be quite challenging. Interestingly, however, it is also the variable degree of degeneration of 5′ss and the surprising heterogeneity existing among human spliceosomal snRNA, which allows for splicing correction using modified exogenous U1 snRNAs.

Overall, the rationale on the use of modified U1 snRNAs to correct splicing defects is as simple as it can be: as 5′ss mutations alter the 5′ss recognition by the endogenous U1 snRNA, exogenous U1 snRNAs may be engineered through complementary base pairing in order to correctly recognize the mutated allele and initiate spliceosome assembly, thus suppressing the mutation effect.

So far, the effects of modified U1 snRNAs have been tested in vitro in a number of cellular platforms from patient-derived cells to model cell lines overexpressing the splicing defects under study, and their potential to either fully or partially correct those mutations was demonstrated for a number of different diseases [5, 6]. Importantly, the application of this sort of modified U1 snRNAs in animal models has also been addressed in recent studies, with a few promising results reported to date [7–10] (*see* **Note 1**).

Globally, mutations affecting constitutive 5′ss represent roughly 8% of all known genetic disease-causing variants. Their pathogenicity derives from the reduced complementarity of the U1 snRNA to the 5′ss. 5′ss mutations mostly result in exon skipping but their effect over splicing may vary. Currently, there are a number of in silico tools that may help predict disease-causing effects, but cDNA analysis remains mandatory for a proper assessment of their consequence over splicing. For example, mutations affecting RNA splicing represent more than 20% of the mutant alleles in Mucopolysaccharidosis type IIIC (MPS IIIC; *HGSNAT* gene), a rare lysosomal storage disorder that causes severe neurodegeneration. Many of these mutations are located in the conserved splice donor or acceptor sites, while few are found in the nearby nucleotides. For three mutations that affect the donor site, we have previously developed different modified U1 snRNAs with compensatory changes that may allow for proper recognition of the mutated 5′ss, in an attempt to rescue the normal splicing process.

A

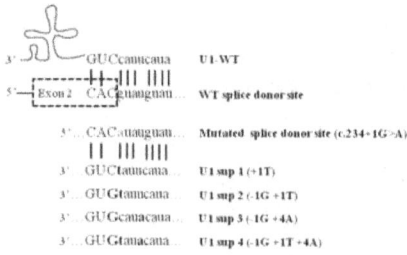

B

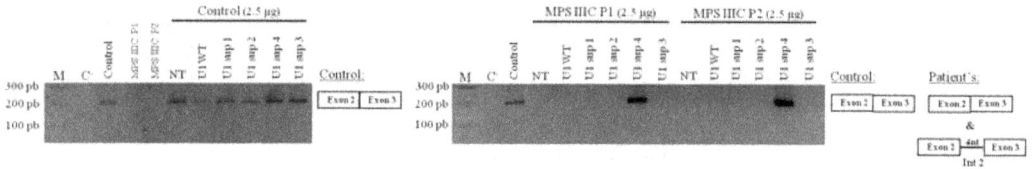

Fig. 1 Modified U1 snRNA therapeutic approach to correct the pathogenic effect of a 5′ splice site mutation on the *HGSNAT* gene. (**a**) Schematic illustration of base pairing between the wild-type U1 (U1-WT) and the 5′ss of wild-type and mutant exon 2 of the *HGSNAT* gene. The mutation position in the 5′ss is marked in grey and it is in italics. The different U1 snRNAs used for the mutated 5′ss of *HGSNAT* (designated as U1-sup, for suppressor) are also shown. The U1 sequence modifications are illustrated in bold. (**b**) RT-PCR analysis of the endogenous splicing pattern of control and MPS IIIC patients derived fibroblasts after transfection with different U1 isoforms. The constitutive splicing of exon 2 of the *HGSNAT gene* was not altered in control fibroblasts after overexpression of U1-WT or any of the modified U1 constructs. In the MPS IIIC patients 1 (MPS IIIC P1) and 2 (MPS IIIC P2), bearing the homozygous mutation c.234+1G>A, only the fully adapted U1 (U1-sup4) resulted in partial correction of exon 2 skipping

For the c.234+1G>A mutation, a totally complementary U1 snRNA allowed for partial correction of exon 2 aberrant splicing in patients' fibroblasts (Fig. 1) [11]. Here, we take advantage of our experience on the development of modified U1 snRNAs to compensate for those *HGSNAT* mutations, to present a practical overview on how to design U1 snRNA molecules for splicing rescue.

In summary, we present an overview of the experimental design for in vitro testing the potential of modified U1 snRNA vectors to correct aberrant splicing caused by 5′ss mutations. Briefly, we show: (a) how to design in silico U1's with different degrees of complementarity to each mutated 5′ss by introducing a number of sequence changes, and (b) how the different U1 vectors harboring those alterations are obtained by site-directed mutagenesis of the original wild-type (WT) human U1 snRNA-harboring pG3U1 vector [12], a derivative of pHU1 [13]. We also describe how these molecules are transfected into patients' fibroblasts and how their effectiveness on splicing redirection can be assessed by post-transfection cDNA analysis and sequencing. Finally, we elaborate on the relevance of further addressing the treatment's effect at protein level.

2 Materials

2.1 Generating Modified U1 snRNA Vectors Adapted to the 5′ss of Interest

1. The *Homo sapiens* U1 snRNA gene sequence is required to design primers for site-directed mutagenesis PCR and can be found in the Ensembl database (ENSG00000104852).

2. The sequence of the 5′ss of interest for splicing rescue can be found in Ensembl or other reference sequence databases (in this particular chapter we used the *Homo sapiens HGSNAT* gene sequence, ENSG00000165102).

3. pG3U1 vector [12] a derivative of pHU1 [13] (*see* **Note 2**).

4. Sense and antisense mutagenic primers.

5. PCR mutagenesis kit.

6. PCR thermocycler.

7. Chemically *Escherichia coli* competent cells (Homemade or commercial; usually are included in the PCR mutagenesis kits).

8. Water bath.

9. Thermomixer.

10. Ice.

11. Super optimal broth with catabolite repression (SOC) medium (commercially available).

12. Luria-Bertani (LB) agar medium (commercially available; sterilize by autoclaving) plates with selection antibiotic (100 µg/mL, ampicillin; *see* **Note 3**).

13. Sterile bacterial cell spreaders.

14. Plasmid DNA miniprep purification kit.

15. LB liquid medium (commercially available; sterilize by autoclaving).

16. Ampicillin.

17. 15 mL conical centrifuge tubes.

18. Sterile tips.

19. Orbital shaking incubator.

20. pG3U1 forward primer (U1-seq Fw—5′ CACGAAG GAGTTCCCGTG 3′).

21. Sterile flasks (1 L).

22. Endotoxin-free maxiprep plasmid DNA purification kit.

23. 40% Glycerol (sterilize by autoclaving).

24. 2 mL polypropylene conical tubes.

2.2 In Vitro Therapeutic Evaluation of Modified U1 snRNA Vectors in Human Fibroblasts

2.2.1 Transfection of Modified U1 snRNA Vectors in Human Fibroblasts

1. Human Dermal Fibroblasts from patients harboring the mutation under analysis (e.g. fibroblasts from patients' with MPS IIIC, carrying the c.234+1G>A mutation in homozigosity) and WT Human Dermal Fibroblasts to use as control.

2. Dulbecco's Modified Eagle's Medium (DMEM) + Glutamax supplemented with 10% Fetal Bovine Serum (FBS), 5% penicillin/streptomycin (PenStrep) antibiotics, and 5% amphotericin B (Fungizone®).

3. Phosphate buffered saline 1x (PBS).

4. Trypsin-EDTA.

5. CO_2 incubator.

6. 15 mL conical centrifuge tubes.

7. Refrigerated centrifuge.

8. Neubauer chamber (hemocytometer).

9. Inverted Microscope.

10. Hand cell counter.

11. T-75 cm^2 cell culture flasks.

12. 6-well cell culture plates.

13. Opti-MEM™ Reduced Serum Medium.

14. Transfection reagent.

15. 1.5 and 2 mL polypropylene conical tubes.

16. Modified U1 snRNA constructs (*see* Subheading 2.1).

2.2.2 Analysis of Splicing Rescue by RT-PCR

1. RNA isolation kit.

2. Refrigerated centrifuge.

3. 1.5 mL polypropylene conical tubes.

4. Spectrophotometer for nucleic acids measurement.

5. cDNA synthesis kit.

6. Taq DNA polymerase.

7. Oligo(dT)$_{18}$ primer mix (if required).

8. Gene-specific primers (e.g. *HGSNAT* primers—Exon 2 Fw: 5′ ACATGCAGAGCTGAAGATGGA 3′; Exon 3 Rv: 5′ GATA GATCCGTGCTGGGTG 3′).

9. Ice.

10. RNase free water.

11. PCR thermocycler.

12. Agarose gel with ethidium bromide for electrophoresis.

13. DNA Ladder (molecular weight size marker).

14. UV transilluminator.

15. Sterile scalpel blades.

16. PCR products purification kits.

3 Methods

3.1 Generating the Modified U1 snRNA Vectors

To design the primers for producing the desired modified human U1 snRNA vectors, it is first necessary to know the sequences of the 5′ss under study, both WT and mutant. Then, it is necessary to analyze the complementarity of those sequences with that of U1 snRNA. Next, several modified U1 snRNA vectors can be designed and constructed to have different complementarities to the target sequences (Fig. 2). To generate those constructs, the plasmid pG3U1 [12] (kindly provided by Prof. Dr. Belén Pérez) a derivative of pHU1 [13], containing the coding sequence of the human U1 can be used as template for site-directed mutagenesis PCR reactions (*see* **Note 2**). Depending on the number of mutations to insert in the U1 snRNA vector sequence, different mutagenic primer pairs need to be designed.

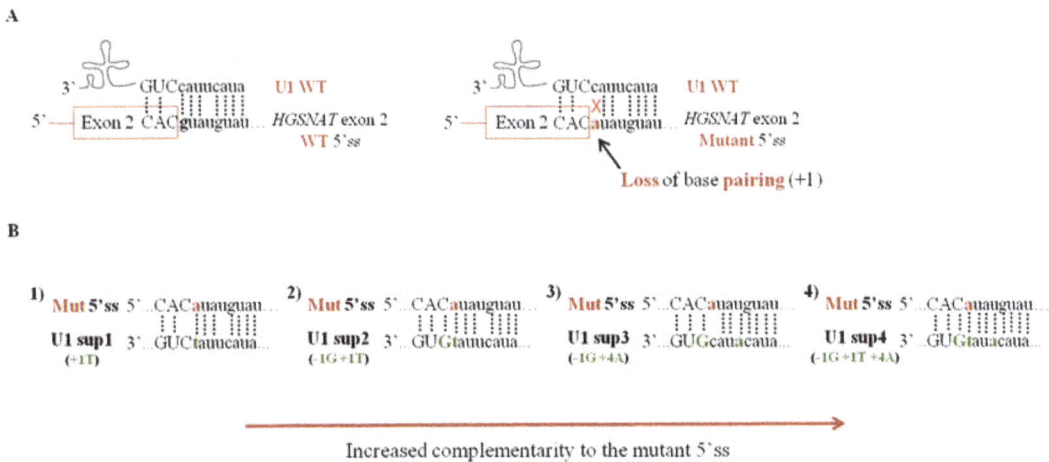

Fig. 2 Design and construction of modified U1 snRNA vectors. (**a**) Schematic representation of base pair interactions between the U1 snRNA and the wild-type and mutant 5′ss of *HGSNAT* exon 2, respectively. (**b**) Illustration of the strategy followed to increase the complementarity of U1 snRNA with the mutated 5′ss of *HGSNAT* gene. U1 complementarity was increased stepwise, and to try to compensate for the *HGSNAT* mutation at +1 position, four different U1-adaptations were designed [U1 sup1 (+1T); U1 sup2 (−1G +1T); U1 sup3 (−1G +4A); U1 sup4 (−1G +1T +4A)]. Upper case letters show exonic nucleotides, whereas the lower case letters denote intronic nucleotides. Base pairing is indicated by vertical lines and its loss by an *X*. The mutant nucleotide is highlighted in red and the changed nucleotides in the U1 sequence are illustrated in green

3.1.1 Engineering
Modified U1 snRNA Vectors
Adapted to the 5′ ss of
Interest

1. According to the different modifications to be introduced in the U1 snRNA vector sequence, design sense and antisense primers with the desired mutation(s) to be introduced by site-directed mutagenesis (*see* **Note 4**).

2. Using the mutagenic primers, perform the site-directed mutagenesis of the WT U1 snRNA vector using the mutagenesis kit (*see* **Note 5**). Briefly, mix the U1 snRNA plasmid DNA (~40 ng) with primers, buffer, dNTPs (according to the kit), apyrogenic water, a High Fidelity Taq polymerase and subject the mixture to recommended PCR conditions from the mutagenesis kit. The number of PCR cycles varies according to the type of the desired mutation(s) (*see* **Note 6**); and the number (n) of min of the PCR extension step depends on the plasmid length, n is calculated as 1 min/kb; *see* **Note 7**. After the PCR reaction is completed, add 1 μL (10 U) of *Dpn*I restriction enzyme to the amplified products and incubate for 1 h at 37 °C to digest the parental dsDNA.

3. Use 1–4 μL of the *Dpn*I treated DNA reaction to transform *E. coli* competent cells. Briefly, thaw on ice a 50 μL aliquot of competent cells and add 1–4 μL of the digested reaction. Swirl the tube gently to mix and incubate on ice for 30 min. In a water bath or dry thermomixer, heat pulse the tube at 42 °C for 45 s and then place the reaction tube on ice for 2 min. Add room temperature SOC medium (5× the volume of competent cells) and incubate for 1 h with shaking at 600 rpm in a dry thermomixer (*see* **Note 8**). After incubation spread the appropriate volume (*see* **Note 9**) of transformation reaction on pre-warmed (37 °C) LB-agar plates containing ampicillin (100 μg/mL) and incubate at 37 °C for 16–18 h (*see* **Note 10**).

4. To obtain plasmid DNA minipreps, prepare minicultures of selected bacterial colonies to allow their growth. Add 3 mL (*see* **Note 11**) of LB medium containing ampicillin (100 μg/ mL) to a 15 mL tube and using a sterilized pipette tip pick a colony and add it into the medium by pipetting up and down (or, simply, place the pipet tip into the medium). Repeat the procedure for 3–5 colonies. Incubate the tubes in an orbital shaking incubator at 220 rpm and 37 °C for 16–18 h. To purify the plasmid DNA prepare DNA minipreps using a plasmid miniprep purification kit (*see* **Note 12**). Select the mutant (s) U1 snRNA plasmid(s) by Sanger sequencing analysis (U1-seq Fw primer) using ~100 ng of purified miniprep.

5. Once the desired modified U1 snRNA construct(s) are selected, propagate them in maxicultures to obtain a high quantity of the modified construct(s) that can be used for transfection. First, prepare a miniculture of each case according to **step 4** (*see* **Note 13**). Then add 100–150 mL of LB medium

containing ampicillin (100 μg/mL) to a sterilized flask(s) (*see* **Note 11**) and innoculate all the bacterial growth from the miniculture(s). Incubate the flask(s) in an orbital shaking incubator at 37 °C and 220 rpm for 16–18 h. Using an endotoxin-free maxiprep plasmid DNA purification kit, maxiprep the plasmid(s) containing the modified U1 snRNA construct(s) and perform its sequencing analysis for validation.

3.2 In Vitro Therapeutic Evaluation of Modified U1 snRNA Vectors in Human Fibroblasts

Even though we must always find a balance between the best possible experimental design and the resources available, adequate controls may never be forgotten. Still, there is a minimum standard for cell culture experiments that must always be met if we want to draw strong conclusions out of them. Therefore, adequate controls to the variables under test should always be included (*see* **Note 14**).

3.2.1 Modified U1 snRNA Vectors Transfection in Human Fibroblasts

1. Grow both WT control and patient fibroblasts in T-75 flasks with DMEM + Glutamax medium supplemented with 10% FBS, 5% antibiotics, and 5% amphotericin B, in an incubator at 37 °C with 95% humidity and 5% CO_2 following standard cell culture procedures.

2. On the day before transfection, detach the cells by trypsinization. Briefly, discard the growth medium and wash cells with 3 mL of PBS buffer. Then, discard the PBS and add 2 mL of trypsin-EDTA. Subsequently, incubate cells with the solution for 5 min at 37 °C. After this period, check in an inverted microscope that cells are detached and add 4 mL of fresh medium to inactivate trypsin-EDTA action.

3. Harvest the cells to a 15 mL tube and centrifuge at $500 \times g$ for 5 min to eliminate any traces of trypsin.

4. Discard the supernatant and resuspend cells in 4 mL of fresh medium.

5. Count cells in suspension with an hemocytometer (Neubauer chamber). Pipette a small volume of cell suspension (approximately 15 μL) to both hemocytometer chambers and count the cells present in all four external quadrants of each chamber by observing it in an inverted microscope. Considering the dimensions of the chamber (1 mm × 1 mm × 0.1 mm), each quadrant has a total volume of 0.1 mm^3, which equals 10^{-4} mL. Therefore, the total number of cells in the original suspension can be calculated with the following equation:

$$N = \frac{\sum n}{8} \times 10^4$$

where N is the total number of cells per milliliter, n is the number of cells counted in each quadrant of the Neubauer chamber and the 10^4 factor allows for the correction of the total number of cells in 1 mL of cell suspension.

6. For modified U1 snRNA vectors transfection, seed a total of $\sim$2.5–3 $\times$ 10^5 fibroblast cells into 6-well plates and grow cells in DMEM + Glutamax medium supplemented with 10% FBS, 5% antibiotics and 5% amphotericin B, in an incubator at 37 °C with 95% humidity and 5% CO_2.

7. On the next day (cells at 80–90% confluence), transfect the cells with quantities between 1 and 3.5 μg of the modified U1 snRNA constructs using a transfection reagent according to the manufacturer's protocol (*see* **Notes 15** and **16**).

8. 24–48 h after transfection, harvest cells by trypsinization. Discard the growth medium of each plate well and wash cells with 1 mL of PBS buffer. Discard the PBS, add 500 μL of trypsin-EDTA to each well and incubate for 5 min at 37 °C. Then, check by microscopy that cells are rounding up and add 1 mL of DMEM + Glutamax medium to inactivate trypsin-EDTA. Harvest cells to 2 mL tubes and centrifuge at 500 $\times$ *g* for 5 min at 4 °C; discard the supernatant; wash cells with 1 mL of PBS buffer and centrifuge again. Proceed to RNA extraction or store the pellet(s) at −80 °C for future use.

3.2.2 Analysis of Splicing Rescue by RT-PCR

1. Extract total RNA from the transfected human fibroblasts using a RNA extraction kit according to the manufacturer's protocol. Then, perform RNA quantification using a spectrophotometer.

2. For reverse transcription, use a cDNA synthesis kit following the manufacturer's protocol, and start with 1–2 μg of total RNA. The cDNA synthesis reaction can be stored at −20 °C or used immediately for PCR amplification.

3. Perform a PCR in standard conditions using a Taq polymerase supplemented with its buffer, dNTPs, gene-specific primers for a final concentration of 0.4 μM each (e.g. *HGSNAT* primers), 2 μL of cDNA, and RNase free water to a final volume of 50 μL.

4. To evaluate the splicing rescue, analyze the amplification products through agarose gel electrophoresis in an agarose gel stained with 5 μL of ethidium bromide (*see* **Note 17**). Choose a DNA ladder according to the size of the amplified band. After separation, visualize the gel using an UV transilluminator. As an example, Fig. 1 shows the results of the partial correction of *HGSNAT* exon 2 splicing after expression of a modified U1 snRNA (totally complementary to the 5′ss of exon 2) in patients' fibroblasts.

5. Assess the identity of the obtained band(s) by sequencing analysis (*see* **Note 18**). For this purpose, purify the PCR products directly with a PCR clean-up kit if there is only one amplified band or when multiple bands are present excise each band from the gel and purify them using a gel band

purification kit. Whatever the case, follow the indications present in the manufacturer's protocol.

6. Subject the purified bands to standard automated sequencing using gene-specific primers for the amplification (e.g. *HGSNAT* primers). Compare the obtained sequence (s) with the reference sequence of the gene of interest (retrieved from the Ensembl database) using the Clustal Omega bioinformatic tool (https://www.ebi.ac.uk/Tools/msa/clustalo/), in order to analyze the effect of the modified U1 snRNA's in rescuing the normal splicing pattern.

3.2.3 Assessment of the Effect of U1 snRNA-Induced Splicing Rescue at Protein Level

While not included in this chapter, for it is case-specific, the effect of modified U1 snRNAs-treatment at protein level is mandatory whenever we want to proceed to in vivo studies in order to address the true therapeutic potential of a given U1 snRNA molecule.

Ideally, as soon as we get an RT-PCR pattern that confirms splicing correction to some extent, and that rescue is confirmed by band excision and Sanger sequencing, the overall effect of that rescue at protein level should also be checked. There is a variety of methods we can choose in order to address this issue, from the direct quantification of enzymatic activity (whenever the gene product under analysis has a catalytic activity) to that of the protein itself (through Western blot).

Usually, the method of choice depends on two major factors: the protein itself and the assays available *in house* to assess it. Virtually every method from Western blot to immunofluorescence may be informative and provide extra support to the conclusions drawn from the RT-PCR. Therefore, as a take-home message, we would recommend that, whenever designing U1 snRNA molecules for splicing rescue, the effect should be checked not only at cDNA level, but also at protein level.

4 Notes

1. This chapter is exclusively focused on mutation-adapted U1 snRNAs. Nevertheless, it is important to refer that there is a novel, second generation, of engineered U1 snRNAs, which may be used for therapeutic purposes: the so-called Exon-Specific U1 snRNAs (ExSpeU1). These ExSpeU1s are complementary to non-conserved sequences downstream of mutant 5′ss. In theory, ExSpeU1 is expected to decrease the potential of off-target effects of U1 snRNA-based therapies, while allowing for a single ExSpeU1 to rescue multiple splicing defects that affect a single exon [4–6].

2. The pG3U1 vector [12] {Susani, 2004, TCIRG1-dependent recessive osteopetrosis: mutation analysis, functional

identification of the splicing defects, and in vitro rescue by U1 snRNA} was used, but the human U1 snRNA sequence can be cloned in other standard expression vector(s).

3. Store LB-agar plates with antibiotics at 4 °C in the dark.

4. Mutagenic primers can be designed using the web-based Quik-Change Primer Design Program, available online at www. agilent.com/store/primerDesignProgram.jsp (we recommend to read the "help" section of the program). However, it is important to take into account a number of considerations:

    (a) both mutagenic primers must contain the desired mutation(s) and anneal to the same sequence on opposite strands of the plasmid;

    (b) each primer should have between 25 and 45 bases in length with a melting temperature (Tm) of $\geq$78 °C;

    (c) the desired mutation(s) should be located in the middle of the primer (~12–15 nucleotides of the correct sequence on both sides);

    (d) the primers should have a minimum GC content of 40% and should terminate in one or more C or G bases;

    (e) the primers do not need to be 5' phosphorylated and purification may either be performed by liquid chromatography (HPLC) or by polyacrylamide gel electrophoresis (PAGE).

5. To modify the pG3U1 we recommend to use the Quik-Change™ II mutagenesis kit (Agilent). However, other site-directed mutagenesis commercial kits can be used. The kit should be chosen according to the plasmid length and the type of mutations to introduce.

6. According to the type of mutation(s) to be inserted in the U1 snRNA WT sequence, the number of PCR cycles varies. For point mutations (1 nucleotide change) use 12 cycles; for single aminoacid changes (3 nucleotides) use 16 cycles and for multiple amino acid deletions or insertions ($\geq$4 nucleotides) use 18 cycles.

7. The number (n) of min of PCR extension step recommended is usually 1 min/kb. However, using the QuikChange™ II mutagenesis kit (Agilent) we usually increment the time for 2 min/kb. For the pG3U1 plasmid length, 8 min should work, but from our experience adding one more min to this step (in this case 9 min for extension) gives the best results.

8. If a thermomixer is not available, follow the site-directed mutagenesis kit manufacturer's recommendations concerning shaking of transformation reactions.

9. The entire volume of transformation reaction can be plated on a single LB-agar plate. However, depending on the transformation efficiency this may originate a huge number of colonies which are then difficult to select. Therefore, we recommend to use more than one plate and spread different volumes to increase the probability to obtain individualized colonies (e.g. 200 and 100 μL). When plaquing lower volumes a small quantity (1:1) of SOC medium can be added to the transformation reaction to dilute and help to spread the transformation product.

10. If colonies cannot be selected immediately, store plate(s) at 4 °C.

11. The total volume of the tube should allow a volume of air that is 5× the volume of LB medium (e.g. 3 mL of LB medium in a 15 mL tube; 5 mL of LB medium in a 25 mL tube, etc.).

12. Before starting the miniprep(s) procedure, a sample of bacterial culture can be preserved in a "glycerolate" for future use. For a final volume of 1 mL, add a part of bacterial culture and a part of sterilized glycerol to a 2 mL tube for a final concentration of ~10–15% of glycerol. Vortex immediately and store at −80 °C.

13. To avoid the need to pick another bacterial colony from an LB-agar plate, the glycerolate(s) (*see* **Note 12**) can be used to prepare a new miniculture. Briefly, defrost the glycerolate on ice, scrape it lightly with a pipette tip or aspirate few microliters and pipet them up and down into a tube containing the desired volume of LB medium and ampicillin (100 μg/mL). Incubate the tube(s) in an orbital shaking incubator as recommended in **step 4** of Subheading 3.1.1.

14. In the transfection experiments here referred (*see* **step 7** of Subheading 3.2.1) we included two negative controls: one where only the transfection reagent was added to the cells and other where the minigene expressing the WT U1 sequence was transfected on cells.

15. For liposome-based transfection of fibroblasts, Lipofectamine® 2000 (Invitrogen) or other commercial lipofection reagent can also be tested. To further increase transfection efficacy, the modified U1's can also be inserted into the cells by the electroporation technique. For both methods we recommend to optimize the amount of transfection reagent according to the quantity of modified U1 and number of cells to transfect.

16. To assess transfection efficiency, transfect fibroblasts with a control plasmid encoding GFP or RFP and monitor fluorescence by microscopy. Also, the cell uptake of the modified U1's can be confirmed by PCR with specific primers (U1 Fw—5′ A TCGAAATTAATACGACTCA 3′ and U1 Rv—5′ CTGGGA AAACCACCTTCGT 3′). Otherwise, clone the WT human U1

snRNA cassette from pG3U1 vector in a plasmid encoding GFP and monitor fluorescence and U1 expression simultaneously.

17. Adjust the agarose gel percentage according to the molecular weight of the target amplified products.

18. In RT-PCR analysis after U1 snRNA's transfection, the size of the amplified band(s) seen on the agarose gel can give an idea of whether the aberrantly spliced exon under study is included in the cDNA or not. However, it is necessary to sequence the amplified band(s) from control and patient fibroblasts treated with the different modified U1 snRNAs, to confirm the correct splicing pattern.

Acknowledgments

The authors would like to acknowledge Prof. Dr. Bele´n P ´e rez (Molecular Biology Department, Faculty of Sciences, University Autonoma of Madrid, Spain) for kindly provide the pG3U1 vector. In addition, the authors would like to acknowledge BioMedCentral, Part of Springer Nature for allowing the reproduction in this chapter of an adapted version of a figure originally published in Matos, L. et al. Therapeutic strategies based on modified U1 snRNAs and chaperones for Sanfilippo C splicing mutations. Orphanet J Rare Dis 9, 180 (2014). This work was partially supported by *Fundação para a Ciênciae Tecno-logia* (FCT) IP (project: FCT/PTDC/BBBBMD/6301/2014) and by *The National MPS Society* (project: 2019DGH1642).

References

1. van der Feltz C, Anthony K, Brilot A, Pomeranz Krummel DA (2012) Architecture of the spliceosome. Biochemistry 51(16): 3321–3333. https://doi.org/10.1021/bi201215r

2. Buratti E, Baralle D (2010) Novel roles of U1 snRNP in alternative splicing regulation. RNA Biol 7(4):412–419

3. Carmel I, Tal S, Vig I, Ast G (2004) Comparative analysis detects dependencies among the 5′ splice-site positions. RNA 10(5):828–840. https://doi.org/10.1261/rna.5196404

4. Pinotti M, Bernardi F, Dal Mas A, Pagani F (2011) RNA-based therapeutic approaches for coagulation factor deficiencies. J Thromb Haemost 9(11):2143–2152. https://doi.org/10.1111/j.1538-7836.2011.04481.x

5. Coutinho MF, Matos L, Santos JI, Alves S (2019) RNA therapeutics: how far have we gone? Adv Exp Med Biol 1157:133–177. https://doi.org/10.1007/978-3-030-19966-1_7

6. Hwu WL, Lee YM, Lee NC (2017) Gene therapy with modified U1 small nuclear RNA. Expert Rev Endocrinol Metab 12(3): 171–175. https://doi.org/10.1080/17446651.2017.1316191

7. Balestra D, Faella A, Margaritis P, Cavallari N, Pagani F, Bernardi F, Arruda VR, Pinotti M (2014) An engineered U1 small nuclear RNA rescues splicing defective coagulation F7 gene expression in mice. J Thromb Haemost 12(2): 177–185. https://doi.org/10.1111/jth.12471

8. Lee NC, Lee YM, Chen PW, Byrne BJ, Hwu WL (2016) Mutation-adapted U1 snRNA corrects a splicing error of the dopa decarboxylase gene. Hum Mol Genet 25(23):5142–5147. https://doi.org/10.1093/hmg/ddw323

9. Lee B, Kim YR, Kim SJ, Goh SH, Kim JH, Oh SK, Baek JI, Kim UK, Lee KY (2019) Modified U1 snRNA and antisense oligonucleotides rescue splice mutations in SLC26A4 that cause hereditary hearing loss. Hum Mutat 40(8): 1172–1180. https://doi.org/10.1002/humu.23774

10. Breuel S, Vorm M, Bräuer AU, Owczarek-Lipska M, Neidhardt J (2019) Combining engineered U1 snRNA and antisense oligonucleotides to improve the treatment of a BBS1 splice site mutation. Mol Ther Nucl Acids 18: 123–130. https://doi.org/10.1016/j.omtn.2019.08.014

11. Matos L, Canals I, Dridi L, Choi Y, Prata MJ, Jordan P, Desviat LR, Pérez B, Pshezhetsky AV, Grinberg D, Alves S, Vilageliu L (2014) Therapeutic strategies based on modified U1 snRNAs and chaperones for Sanfilippo C splicing mutations. Orphanet J Rare Dis 9:180. https://doi.org/10.1186/s13023-014-0180-y

12. Susani L, Pangrazio A, Sobacchi C, Taranta A, Mortier G, Savarirayan R, Villa A, Orchard P, Vezzoni P, Albertini A, Frattini A, Pagani F (2004) TCIRG1-dependent recessive osteopetrosis: mutation analysis, functional identification of the splicing defects, and in vitro rescue by U1 snRNA. Hum Mutat 24(3):225–235. https://doi.org/10.1002/humu.20076

13. Lund E, Dahlberg JE (1984) True genes for human U1 small nuclear RNA. Copy number, polymorphism, and methylation. J Biol Chem 259(3):2013–2021

9

Eye on a Dish Models to Evaluate Splicing Modulation

Kwan-Leong Hau, Amelia Lane, Rosellina Guarascio and Michael E. Cheetham

Abstract

Inherited retinal dystrophies, such as Leber congenital amaurosis, Stargardt disease, and retinitis pigmentosa, are characterized by photoreceptor dysfunction and death and currently have few treatment options. Recent technological advances in induced pluripotent stem cell (iPSC) technology and differentiation methods mean that human photoreceptors can now be studied in vitro. For example, retinal organoids provide a platform to study the development of the human retina and mechanisms of diseases in the dish, as well as being a potential source for cell transplantation. Here, we describe differentiation protocols for 3D cultures that produce retinal organoids containing photoreceptors with rudimentary outer segments. These protocols can be used as a model to understand retinal disease mechanisms and test potential therapies, including antisense oligonucleotides (AONs) to alter gene expression or RNA processing. This "retina in a dish" model is well suited for use with AONs, as the organoids recapitulate patient mutations in the correct genomic and cellular context, to test potential efficacy and examine off-target effects on the translational path to the clinic.

Keywords Retinal organoids, Induced pluripotent stem cells, Differentiation, 3D culture, Retinal degeneration, Photoreceptor, Retina in a dish

1 Introduction

The dysfunction and death of photoreceptor cells are associated with inherited retinal diseases (IRDs), which are a major cause of blindness. The lack of effective treatment to prevent loss of photoreceptors means these diseases are currently irreversible. Recent progress in the differentiation of stem cells to retinal cells has enabled the generation of functional retinal organoids in vitro or a "retina in a dish" [1–4]. By recapitulating the retina from patient-derived induced pluripotent stem cells (iPSC), retinal organoids offer a platform for developing therapeutic treatments and modeling patient disease [5, 6].

A dynamic and complex microenvironment is involved in eye development, including direct and indirect cell–cell interaction and specific signaling regulation in different stages of development

[7]. Because of this complex microenvironment, retinal organoids have the potential to develop a more mature retina than photoreceptors differentiated in 2D conditions only. Several studies have shown that with defined culture conditions, embryonic stem cells (ESC), and iPSC can be differentiated into retinal organoids in a 3D environment, producing a laminated retina that mimics the in vivo human retina [2, 3, 8]. In addition to recapitulating the structure of native eye development, rudimentary disorganized outer segments can be observed in photoreceptors from retinal organoids.

In this chapter, we describe three different methods to differentiate iPSC to retinal organoids in 3D. Retinal organoids generated from these protocols are well laminated with photoreceptors in their outer layer and develop rudimentary outer segments. Importantly, they also recapitulate photoreceptor mRNA processing and the exquisite pattern of alternative splicing they present [9–11]. This makes retinal organoids ideal for studying aberrant splicing events associated with patient variants in several forms of IRDs [9, 12]. Furthermore, they can then be applied to the development of antisense oligonucleotides (AONs) as potential treatments [9, 13].

2 Materials

2.1 General Materials

1. U-bottom ultra-low 96-well plate.
2. 25-well plate low attachment.
3. 6-well plate.
4. Ultra-low adhesion 6-well plate.
5. Crescent knife.
6. Retinoic acid (RA).

2.2 iPSC Culture

1. Essential 8 Flex medium.
2. Geltrex.
3. Cell dissociation buffer (ThermoFisher).

2.3 EB Suspension Protocol

1. EB2 base medium: GMEM, 20% Knock-Out Serum, 1% Sodium pyruvate, 1% NEAA, and 110 μM 2-Mercaptoethanol (*see* **Note 1**).
2. NR media (NRM): DMEM/F12, 1% N_2 Supplement, 10% FBS, and 1% NEAA (*see* **Note 1**).
3. V-bottom 96-well plate.
4. IWR-1e (Wnt inhibitor).
5. Rock inhibitor (Y-27632).

6. Matrigel (growth factor reduced).

7. Hedgehog smoothened agonist (SAG).

8. TrypLE.

2.4 EB Adherent Protocol

1. Neural induction medium (NIM): DMEM/F-12 (1:1), 1% N_2 supplement, 1% NEAA, and Heparin 2 μg/ml (*see* **Note 1**).

2. Retinal Differentiation Medium (RDM): DMEM/F12 (3:1), 2% B27 (without Vitamin A), 1% NEAA, and 1% Pen/Step (*see* **Note 1**).

3. Neural Retina Maturation Medium 1 (RMM1): DMEM/F12 (3:1), 2% B27 (without vitamin A), 1% NEAA, 1% Pen/Strep, 10% FBS, 100 μM Taurine, and 1% Glutamax (*see* **Note 1**).

4. Neural Retina Maturation Medium 2 (RMM2): DMEM/F12 (3:1), 1% N2, 1% NEAA, 1% Pen/Strep, 10% FBS, 100 μM Taurine, and 1% Glutamax (*see* **Note 1**).

5. Blebbistatin.

2.5 Non-EB Adherent Protocol

1. Essential 6 medium (ThermoFisher).

2. Neural induction Medium (NIM): Advanced DMEM/F12, 1% N_2 supplement, 1% NEAA, 1% Glutamax, and 1% Pen/Strep (*see* **Note 1**).

3. Retinal Differentiation Media (RDM): DMEM/F12 (3:1), 1% Pen/Strep, 1% NEAA, and 2% B27 (*see* **Note 1**).

4. Neural Retina Maturation Medium 1 (RMM1): DMEM/F12 (3:1), 1% Pen/Strep, 2% B27, 10% FBS, 100 μM Taurine, 1% NEAA, and 1% Glutamax (*see* **Note 1**).

5. Neural Retina Maturation Medium 2 (RMM2): DMEM/F12 (3:1), 1% Pen/Strep, 2% B27 (without vitamin A), 1% N_2, 10% FBS, 100 μM Taurine, 1% NEAA, and 1% Glutamax (*see* **Note 1**).

2.6 RNA Extraction

1. RNA mini kit.

2. PBS.

3. Micropestle.

3 Methods

iPSC are maintained with Essential 8 Flex (E8F) in Geltrex coated 6-well plates (*see* **Note 2**). Once they reach 70% confluence, iPSC are treated with 500 μl cell dissociation buffer for 2 min at 37 °C in the incubator. After the incubation, remove the cell dissociation buffer and add 1 ml of E8F into a well. Use 1 ml tip scraping the well to collect iPSC in small clumps. Cell clumps are collected and transferred into a new Geltrex-coated plate with 1 ml tip. Medium is changed every other day, and iPSC can be double-fed with 4 ml E8F to cover the weekend (*see* **Note 3**).

3.1 EB Suspension Protocol

This protocol is adapted from the method initially described by Sasai and colleagues [3].

1. Maintain iPSC in a 6-well plate as described earlier. Use 1 ml TrypLE to disperse cells into single cells. Collect the cell pellet after centrifugation at $300 \times g$ for 5 min and resuspend in 2 ml E8F with 10 μM Rock inhibitor (ROCKi).

2. Place 10,000 cells per well with 100 μl E8F with ROCKi in a V-bottom low attachment 96-well plate.

3. The next day, add 100 μl E8F with ROCKi (Day 1).

4. Change half medium, 100 μl, with EB2 with 10 μM ROCKi, 3 μM IWRe-1, and 2% Matrigel twice a week until Day 12.

5. Change half medium, 100 μl, with EB2 with 3 μM IWRe-1, 2% Matrigel, 10% FBS, and 100 nM SAG twice a week until Day 18.

6. Transfer cells into U-bottom ultra-low attachment 96-well plates, and the medium is switched to NRM supplemented with 0.5 μM RA from Day 20 until Day 100. Change medium three times a week (Fig. 1) (*see* **Note 4**).

7. From Day 100, select and maintain laminated organoids in 25-well plates in NRM with no RA till collection day (Fig. 1) (*see* **Note 5**).

3.2 EB Adherent Protocol

This protocol is adapted from the method initially described by Canto-Soler and colleagues [2].

1. Collect the iPSC clusters, as described earlier, from three confluent wells (or one T25 flask) with E8F + 10 μM Blebbistatin (*see* **Note 6**) and transfer the cell clumps into three wells (2 ml per well) of ultra-low adhesion 6-well plate to form the embryoid bodies (EB).

2. After 24 h (Day 1), use a 10-ml pipette to collect the EB into 15-ml falcon and centrifuge at $110 \times g$ for 2 min. Remove the supernatant and collect the EB with 6 ml medium of 75% E8F + 10 μM Blebbistatin and 25% neural induction medium (NIM). Transfer the EB back to the wells, 2 ml in each well (*see* **Note 7**).

3. With the same technique described in **step 2**, change the medium to 50% E8F + 10 μM Blebbistatin and 50% NIM on Day 2 and 100% NIM on Day 3 and Day 5 (*see* **Note 8**).

4. On Day 7, transfer EB from three wells to six wells of Geltrex-coated 6-well plate in NIM, 4 ml in each well. Gently mix the

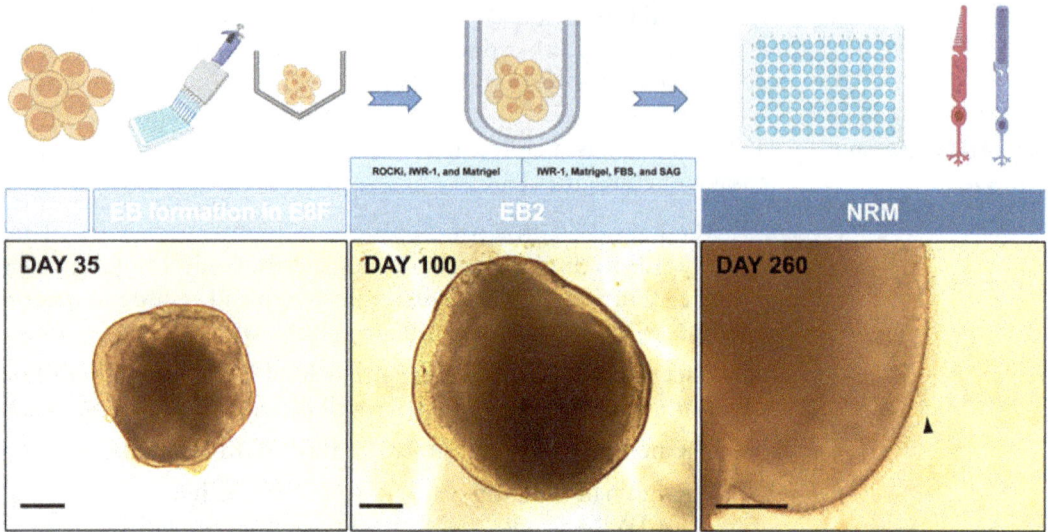

Fig. 1 EB suspension protocol. Top row: Schematic diagram of EB suspension protocol steps and media. Lower row: Representative organoids at different stages of differentiation are shown. Visible lamination can be observed at approximately Day 35 and good organoids can maintain the lamination and mature during differentiation to form an outer nuclear layer of photoreceptors with inner segment and outer segment (which can be seen by the "brush border," arrowhead). Scale bar is 250 μm. The cartoon images are made with BioRender

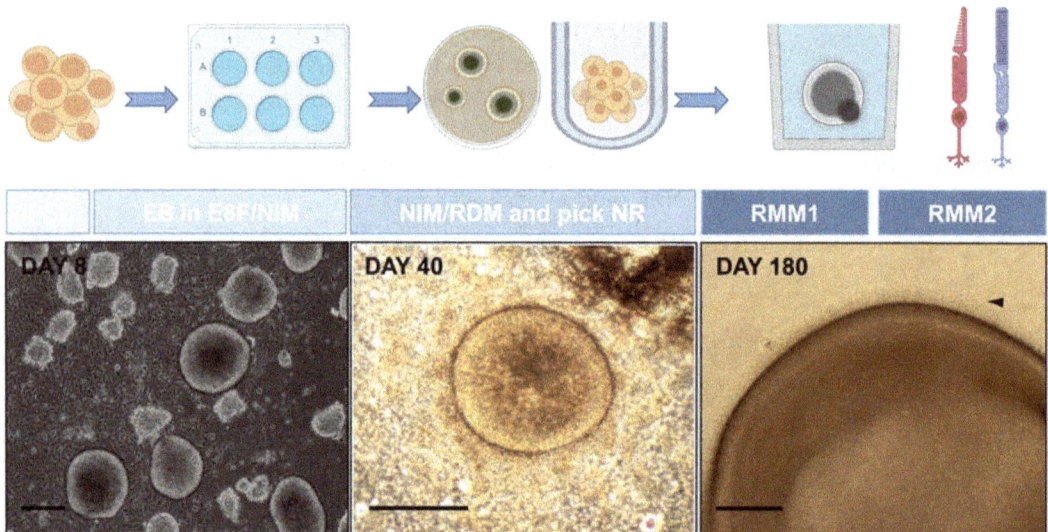

Fig. 2 EB adherent protocol. Top row: Schematic diagram of EB adherent protocol steps and media. Lower row: Representative images at different stages of differentiation are shown. EB formation in suspension is followed by attachment to Geltrex-coated wells and formation of NR. Picked NR successfully mature through the differentiation form an outer nuclear layer of photoreceptors with inner segment and outer segment (which can be seen by the "brush border," arrowhead). Scale bar is 250 μm. The cartoon images are made with BioRender

medium in the wells to let the EB equally distributed in the wells (Fig. 2) (*see* **Note 2**).

5. Change 4 ml NIM medium twice a week until Day 15.

6. Feed the cells daily with retinal differentiation medium (RDM) from Day 16 until neural retina (NR) domains [9] are formed (Fig. 2).

7. Pick individual NR mechanically with a crescent knife using an inverted brightfield microscope (EvosXL Core) in the safety cabinet in between Day 28 and Day 35 and culture in suspension in U-bottom ultra-low 96-well plates with RDM, one NR per well. Change medium three times a week (*see* **Notes 4** and **11**).

8. Switch medium to RMM1 from Day 42. Of note, 1 µM RA is introduced from Day 63.

9. Switch medium to RMM2 with 0.5 µM RA from Day 90 and RMM2 only from Day 100.

10. On Day 100, laminated retinal organoids can be observed with a microscope and are selected and transferred to 25-well plates with 1 ml medium (*see* **Note 5**).

11. Maintain retinal organoids in RMM2 until collection day.

3.3 Non-EB Adherent Protocol

This protocol is adapted from the method initially described by Ali and colleagues [8].

1. iPSC are maintained in 6-well plates as described earlier. E8F is switched to Essential 6 medium for 2 days when the cells reach 90–100% confluence (*see* **Note 9**).

2. Introduce 4 ml NIM from Day 3 and change the medium three times a week until the NR are formed (Fig. 3) (*see* **Note 10**).

3. Pick NR, as described in EB Adherent Protocol (*see* Subheading 3.2, **step** 7), and maintain in U-bottom ultra-low attachment 96-well plates with RDM up to 1 week (*see* **Notes 4** and **11**).

4. Switch medium from RDM to RMM1 for 4 weeks.

5. After 4 weeks with RMM1 only, introduce 1 µM RA in RMM1 for 2 weeks.

6. Switch medium to RMM2 with 0.5 µM RA until Day 100.

7. Visually confirmed laminated organoids are transferred to 25-well plates and maintained in RMM2 without B27 until collection day (Fig. 3) (*see* **Note 12**).

3.4 AON Treatment of Organoids

1. Mature organoids are generated from protocols described in the above sections.

2. Dilute AONs into working concentration (e.g., 0.1–10 µM) with culture medium, depending on the methods (*see* **Note 13**).

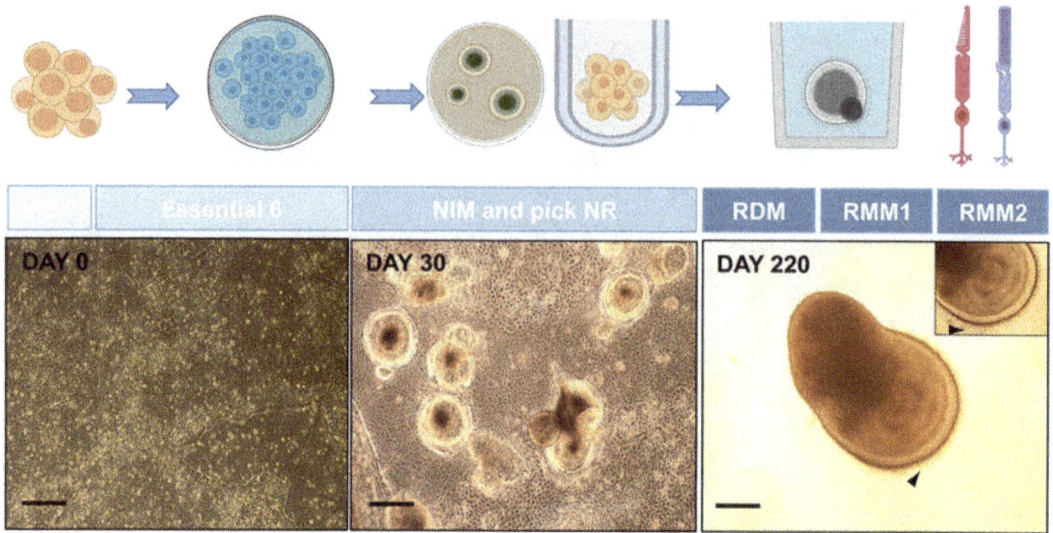

Fig. 3 Non-EB adherent protocol. Top row: Schematic diagram of non-EB adherent protocol steps and media. Lower row: Representative images at different stages of differentiation are shown. NR are formed in NIM medium, and picked NR cultured in suspension going through differentiation mature to form organoids with an outer nuclear layer of photoreceptors with inner segment and outer segment projecting outwards (which can be seen by the "brush border," arrowhead). Scale bar is 250 μm. The cartoon images are made with BioRender

3. Remove the medium and treat organoids with media containing AONs (*see* **Note 14**).

4. Treat organoids with AONs two times a week with a full change of the medium containing the AON (*see* **Note 15**).

5. On the collection day, transfer the organoids into 1.5-ml microcentrifuge tubes with 1 ml PBS individually.

6. Remove PBS and keep the microcentrifuge tube on dry ice for 10 min.

7. RNA can be extracted immediately or the samples can be stored at −80 °C (*see* **Note 16**).

3.5 RNA Extraction from Organoids

1. Samples are prepared as described in the previous step.

2. Homogenize organoids individually with micropestle in the microcentrifuge tube.

3. Add lysis buffer from RNA mini kit, in this case from Qiagen, and homogenize organoids again (*see* **Note 17**).

4. Follow the instruction of RNA extraction kit to finish RNA extraction and cDNA can then be synthesized.

3.6 Read-out

These methods can be used to produce laminated retinal organoids for the study of RNA processing and morphological changes associated with genomic variants and their potential correction with

AONs. The assays used for downstream analyses are dependent on the specific questions being asked. Routine analyses would usually involve RT-PCR and qPCR, but the organoids are also amenable to RNAseq, single-cell sorting, next-generation sequencing, or long-range sequencing. This can provide a unique insight into human photoreceptor splicing and its manipulation for discovery science or therapeutic benefit.

4 Notes

1. General. Once supplemented, the complete medium is stable for up to 2 weeks when stored in the fridge at 4 °C. Freshly made medium can be aliquoted and stored in the freezer at −20 °C for longer storage.

2. Geltrex from stock solution is diluted 50 times in DMEM/F12 medium and 1 ml diluted Geltrex is used to coat a well in a 6-well plate. Plate is coated at 37 °C in the incubator for an hour. EB will attach in the wells from this step.

3. General. Different iPSC clones might have different efficiency of differentiation using these methods. It is recommended to start with at least two of the protocols to test which method is more efficient for that specific clone. The retinal identity and correct lamination of organoids produced by any of the three described methods can be predicted by careful visual inspection under a microscope, but it must be verified by expression of mature retinal markers (e.g., recoverin, cone arrestin, rhodopsin, LM opsin) by immunofluorescence staining and/or gene expression assays.

4. Medium is changed three times a week for 96-well plates and two times a week for 25-well plates.

5. Cut the end of a 1-ml pipette tip off to transfer organoids from 96-well plates to 25-well plates. One retinal organoid per well of 25-well plates. More than one organoid in a well might cause them to merge together.

6. Three wells from a 6-well plate or one T25 flask are optimized conditions we use, but this vary depending on the size of the clumps and confluence of the iPSC. So this step may need to be optimized in each lab.

7. Be gentle while collecting and transferring EB from and into wells. Avoid breaking the EB into single cells.

8. Gamm and colleagues reported that a single dose of BMP4 at day 6 of differentiation, followed by one-half media changes every 3 days until day 16, improved NR production [14].

9. iPSCs need to reach almost 100% confluence, this is crucial for non-EB adherent protocol. Lower density might cause cell death and failure of the protocol. Essential 6 medium is changed daily.

10. For weekend feeding, 6 ml of NIM is used on Friday instead of 4 ml. NR are usually formed between week 4 and week 6.

11. At the NR picking step, it is recommended to pick as many as possible (or needed) to increase the number of mature organoids. Between 50% and 90% of the NR picked will not make it to mature laminated retinal organoids (dependent on cell line). Some fail to form organized neuroepithelium in suspension and some collapse in a later stage forming a ball of neuro-retinal rosettes that will not develop the full outer and inner retinal layers. It is necessary to account for this when designing experiments.

12. We find media without B27 from this stage may improve the organization of the inner retinal cell layers in organoids.

13. 0.1–10 μM is the concentration range that we have tested for AONs (with phosphorothioate backbone and either 2′-O-methyl or 2′-O-methoxyethyl modifications). The working concentration might vary with different AONs, as this is empirical.

14. Gymnotic treatment with phosphorothioate backbone AONs is effective for retinal organoids. Addition of 6 μM EndoPorter will assist morpholino uptake. To treat the organoids, 200 μl of total volume is used in 96-well plates and 1 ml in 25-well plates.

15. Treatment time is empirical and will depend on the specific target or assay being used. We have used treatment times between 72 h and 4 weeks.

16. Nonsense mediated decay can be inhibited with emetine prior to sample collection, if it is suspected this is affecting the detection of aberrant transcripts.

17. A 30-gauge needle can be used to help homogenize organoids. 200–500 ng of RNA can be extracted per organoid.

Acknowledgments

This work is supported by Wellcome Trust, Fight for Sight, Foundation Fighting Blindness, Retina UK, Moorfields Eye Charity and NC3Rs. We would like to thank the other members of the Cheetham, Hardcastle, and van der Spuy groups past and present for their support, encouragement and help in iPSC and organoid maintenance. We would also like to thank Anai Gonzalez-Cordero for advice on the non-EB adherent protocol.

References

1. Gonzalez-Cordero A, West EL, Pearson RA, Duran Y, Carvalho LS, Chu CJ, Naeem A, Blackford SJI, Georgiadis A, Lakowski J, Hubank M, Smith AJ, Bainbridge JWB, Sowden JC, Ali RR (2013) Photoreceptor precursors derived from three-dimensional embryonic stem cell cultures integrate and mature within adult degenerate retina. Nat Biotechnol 31(8):741–747. https://doi.org/10.1038/nbt.2643

2. Zhong X, Gutierrez C, Xue T, Hampton C, Vergara MN, Cao LH, Peters A, Park TS, Zambidis ET, Meyer JS, Gamm DM, Yau KW, Canto-Soler MV (2014) Generation of three-dimensional retinal tissue with functional photoreceptors from human iPSCs. Nat Commun 5:4047. https://doi.org/10.1038/ncomms5047

3. Nakano T, Ando S, Takata N, Kawada M, Muguruma K, Sekiguchi K, Saito K, Yonemura S, Eiraku M, Sasai Y (2012) Self-formation of optic cups and storable stratified neural retina from human ESCs. Cell Stem Cell 10(6):771–785. https://doi.org/10.1016/j.stem.2012.05.009

4. Meyer JS, Shearer RL, Capowski EE, Wright LS, Wallace KA, McMillan EL, Zhang SC, Gamm DM (2009) Modeling early retinal development with human embryonic and induced pluripotent stem cells. Proc Natl Acad Sci U S A 106(39):16698–16703. https://doi.org/10.1073/pnas.0905245106

5. Gamm DM, Phillips MJ, Singh R (2013) Modeling retinal degenerative diseases with human iPS-derived cells: current status and future implications. Expert Rev Ophthalmol 8(3):213–216. https://doi.org/10.1586/eop.13.14

6. Sasai Y (2013) Next-generation regenerative medicine: organogenesis from stem cells in 3D culture. Cell Stem Cell 12(5):520–530. https://doi.org/10.1016/j.stem.2013.04.009

7. Bassett EA, Wallace VA (2012) Cell fate determination in the vertebrate retina. Trends Neurosci 35(9):565–573. https://doi.org/10.1016/j.tins.2012.05.004

8. Gonzalez-Cordero A, Kruczek K, Naeem A, Fernando M, Kloc M, Ribeiro J, Goh D, Duran Y, Blackford SJI, Abelleira-Hervas L, Sampson RD, Shum IO, Branch MJ, Gardner PJ, Sowden JC, Bainbridge JWB, Smith AJ, West EL, Pearson RA, Ali RR (2017) Recapitulation of human retinal development from human pluripotent stem cells generates transplantable populations of cone photoreceptors. Stem Cell Rep 9(3):820–837. https://doi.org/10.1016/j.stemcr.2017.07.022

9. Parfitt DA, Lane A, Ramsden CM, Carr AJ, Munro PM, Jovanovic K, Schwarz N, Kanuga N, Muthiah MN, Hull S, Gallo JM, da Cruz L, Moore AT, Hardcastle AJ, Coffey PJ, Cheetham ME (2016) Identification and correction of mechanisms underlying inherited blindness in human iPSC-derived optic cups. Cell Stem Cell 18(6):769–781. https://doi.org/10.1016/j.stem.2016.03.021

10. Ling JP, Wilks C, Charles R, Leavey PJ, Ghosh D, Jiang L, Santiago CP, Pang B, Venkataraman A, Clark BS, Nellore A, Langmead B, Blackshaw S (2020) ASCOT identifies key regulators of neuronal subtype-specific splicing. Nat Commun 11(1):137. https://doi.org/10.1038/s41467-019-14020-5

11. Kim S, Lowe A, Dharmat R, Lee S, Owen LA, Wang J, Shakoor A, Li Y, Morgan DJ, Hejazi AA, Cvekl A, DeAngelis MM, Zhou ZJ, Chen R, Liu W (2019) Generation, transcriptome profiling, and functional validation of cone-rich human retinal organoids. Proc Natl Acad Sci U S A 116(22):10824–10833. https://doi.org/10.1073/pnas.1901572116

12. Buskin A, Zhu L, Chichagova V, Basu B, Mozaffari-Jovin S, Dolan D, Droop A, Collin J, Bronstein R, Mehrotra S, Farkas M, Hilgen G, White K, Pan KT, Treumann A, Hallam D, Bialas K, Chung G, Mellough C, Ding Y, Krasnogor N, Przyborski S, Zwolinski S, Al-Aama J, Alharthi S, Xu Y, Wheway G, Szymanska K, McKibbin M, Inglehearn CF, Elliott DJ, Lindsay S, Ali RR, Steel DH, Armstrong L, Sernagor E, Urlaub H, Pierce E, Luhrmann R, Grellscheid SN, Johnson CA, Lako M (2018) Disrupted alternative splicing for genes implicated in splicing and ciliogenesis causes PRPF31 retinitis pigmentosa. Nat Commun 9(1):4234. https://doi.org/10.1038/s41467-018-06448-y

13. Dulla K, Aguila M, Lane A, Jovanovic K, Parfitt DA, Schulkens I, Chan HL, Schmidt I, Beumer W, Vorthoren L, Collin RWJ, Garanto A, Duijkers L, Brugulat-Panes A, Semo M, Vugler AA, Biasutto P, Adamson P,

Cheetham ME (2018) Splice-modulating oligonucleotide QR-110 restores CEP290 mRNA and function in human c.2991+1655A>G LCA10 models. Mol Ther Nucleic Acids 12:730–740. https://doi.org/10.1016/j.omtn.2018.07.010

14. Capowski EE, Samimi K, Mayerl SJ, Phillips MJ, Pinilla I, Howden SE, Saha J, Jansen AD, Edwards KL, Jager LD, Barlow K, Valiauga R, Erlichman Z, Hagstrom A, Sinha D, Sluch VM, Chamling X, Zack DJ, Skala MC, Gamm DM (2019) Reproducibility and staging of 3D human retinal organoids across multiple pluripotent stem cell lines. Development 146(1). https://doi.org/10.1242/dev.171686

10

In Vitro Delivery of PMOs in Myoblasts by Electroporation

Remko Goossens and Annemieke Aartsma-Rus

Abstract

Antisense oligonucleotides (AONs) are small synthetic molecules of therapeutic interest for a variety of human disease. Their ability to bind mRNA and affect its splicing gives AONs potential use for exon skipping therapies aimed at restoring the dystrophin transcript reading frame for Duchenne muscular dystrophy (DMD) patients. The neutrally charged phosphorodiamidate morpholino oligomers (PMOs) are a stable and relatively nontoxic AON modification. To assess exon skipping efficiency in vitro, it is important to deliver them to target cells. Here, we describe a method for the delivery of PMOs to myoblasts by electroporation. The described protocol for the Amaxa 4D X unit nucleofector system allows efficient processing of 16 samples in one nucleocuvette strip, aiding in high-throughput PMO efficacy screens.

Key words AON, PMO, Electroporation, Nucleofection, Myocytes, Duchenne muscular dystrophy, DMD

1 Introduction

Antisense oligonucleotides (AONs) are versatile, powerful tools for the potential treatment of a variety of diseases. AONs are short synthetic oligonucleotides consisting of modified DNA or RNA nucleic acid analogs. AONs can be exploited in multiple ways, including the modulation of splicing. Here, AONs bind to the unspliced mRNA and mask splice sites or exonic splice enhancer or silencer sites, resulting in an exon being ex- or included in the mature mRNA. Examples of such AONs are eteplirsen, golodirsen, viltolarsen, and casimersen to treat Duchenne muscular dystrophy (DMD) patients with eligible mutations, and nusinersen to treat spinal muscular atrophy (SMA). For DMD, antisense-mediated exon skipping is used to restore the reading frame of the dystrophin (*DMD*) transcript, allowing the production of an internally deleted partially functional dystrophin protein [1]. This approach is mutation specific. Currently, for Duchenne, four AONs have been approved by the Food and Drug Administration (FDA, USA), which induce skipping of exon 51 (eteplirsen), exon 53 (golodirsen and viltolarsen) or exon 45 (casimersen).

Various different AON chemistries have been developed, such as 2'-O-methyl phosphorothioate (2'O-MePS), 2'-O-methoxyethyl phosphorothioate (2'-MOE-PS), and phosphorodiamidate morpholino oligomer (PMO) [2]. These different chemistries have unique chemical properties and were developed to enhance stability, solubility, and cellular uptake of AONs [2]. AONs are sometimes covalently conjugated to other molecules, in an attempt to improve their uptake by target tissues after systemic delivery [3]. For treatment of SMA, intrathecal injection of the AON into the cerebrospinal fluid (CSF) leads to efficient uptake by neurons and other cells in the nervous system with a long half-life [4]. However, in DMD all of the >700 different skeletal muscles are affected, and as such systemic delivery of AONs is required, which currently involves weekly intravenous infusions [5].

DMD consists of 79 exons, and there is a wide variety of unique patient mutations [6, 7], with a mutation hotspot spanning exon 45 through 53 [6, 8]. Approximately 55% of total *DMD* mutations causative for Duchenne would be eligible for some form of exon skipping therapy [7]. While skipping certain exons is applicable to larger groups of patients, it is crucial to skip also additional exons, which individually apply to small groups of patients, to increase the general applicability of this approach to as many patients as possible. To optimally design AONs for most of the *DMD* exons, it is important to have the ability for high-throughput screening of AON exon skipping efficacy. To perform reliable initial testing of AONs in vitro, it is essential to establish a reproducible, efficient means of delivery to a target cell. This can be achieved using electroporation of immortalized muscle cells [9]. Using immortalized cells has the advantage of theoretically unlimited proliferation, so large amounts of cells with homogenous characteristics can be generated. Primary cell sources are finite, and each new donor will have to be validated for reproducibility, which hampers screening potential when a large number of different AONs are to be tested simultaneously. Furthermore, it is our experience that the capacity of primary cultures to differentiate into myotubes declines with advanced passages.

Unlike 2'O-MePS AONs and dsDNA (e.g., plasmids), PMO AONs are neutrally charged, impeding the delivery by cationic lipid transfection systems. An alternative method for delivery of PMOs to mammalian cells is electroporation [9]. Electroporation relies on the formation of pores in the cell membrane by the application of an electric pulse through the transfection medium, mediating delivery of the particle of interest. The pore-forming pulse either serves to simultaneously deliver a charged molecule or is followed by a dedicated secondary delivery pulse. While electroporation efficiency is aided by active mobilization of charged molecules of interest into the cells by the electric current applied, the pore formation itself is already able to allow passive entry of inert molecules such as PMOs, albeit potentially at a lower efficiency. In fact, efficiency of in vitro PMO electroporation is relatively high when compared to other methods such as gymnosis and calcium-enriched medium (CEM) [10].

Classic electroporation is performed with cells in suspension, requiring dissociation of adherent cells from the culture vessel prior to the procedure. For studying *DMD* exon skipping, a cell line which expresses full-length *DMD* (Dp427m) is required. Alternatively, if available, a patient-derived cell line with a specific mutation can be used where skipping of a specific exon restores dystrophin production. However, as *DMD* expression in proliferating myoblasts is very low, myoblasts need to differentiate after electroporation with the Amaxa 4D X unit to form mature myotubes, expressing higher levels of *DMD*. Novel electroporation techniques also allow for electroporation of adherent cell layers, which can be advantageous when working with cells that grow slowly or have limited proliferation capacity upon differentiation, such as mature myotubes or neuronal cells. These methods, just like other electroporation techniques, require thorough characterization and optimization of conditions for maximum efficiency and are not included in this chapter.

In this chapter, we provide a protocol for the delivery of PMOs to immortalised myoblasts by electroporation with the Lonza Amaxa 4D-nucleofector X unit, and their subsequent differentiation to *DMD* expressing myotubes. The procedure for further sample processing and analysis of the skipped transcript by end-point reverse transcription polymerase chain reaction (RT-PCR) and quantitative PCR (RT-qPCR) are also outlined.

2 Materials

2.1 Cell Culture

1. Immortalized myoblasts: 0.5×10^6 up to 1×10^6 cells per reaction (for transfer to a 6-well plate) (*see* **Note 1**).

2. Culture medium (proliferation): For myoblasts either: F10 Nutrient mix (nutmix) medium supplemented with 20% fetal bovine serum (FBS), 1% PenStrep, 10 ng/mL rhFGF and 1 mM dexamethasone, **or** Skeletal Muscle Cell Growth Medium (SMCGM) supplemented with 15% FBS and 50 μg/mL gentamicin (*see* **Note 2**).

3. Culture medium (resuspension): F10 Nutmix + 20% FBS + 1% PenStrep.

4. Culture medium (differentiation): DMEM (4.5 g/L glucose) + 2% FBS **or** 2% knockout serum replacement (KOSR) + 50 μg/mL Gentamicin **or** 1% PenStrep (*see* **Note 2**).

5. Trypsin-EDTA (0.05%).

6. Dulbecco's phosphate-buffered saline (dPBS (-MgCl₂, -CaCl₂)).

7. Culture vessels (e.g., T182 flask and 6-well plates).

2.2 Electroporation

1. Lonza 4D nucleofector core unit.

2. Lonza 4D nucleofector X-unit for cells in suspension (*see* **Note 3**).

3. Suitable 4D-nucleofector X kit (e.g., primary cell optimization kit) containing 16-well nucleofector strips and nucleofection buffer.

4. PMOs at 1 mM concentration in saline, 0.2-μm filter sterilized (*see* **Notes 4 and 5**).

2.3 RNA Isolation

1. TRI-reagent (e.g., TRIsure).

2. Chloroform.

3. 2-Propanol.

4. 70% Ethanol (EtOH).

5. RNase-free water (DEPC treated).

2.4 cDNA Synthesis

1. Random hexamer primers (N6) (20 ng/μL).

2. dNTP mix (10 mM each).

3. 5× reverse transcriptase (RT) reaction buffer.

4. Reverse transcriptase enzyme.

5. RNase inhibitor.

6. RNase-free water (DEPC treated).

2.5 RT-PCR Analysis of Skipping Efficiency

1. cDNA generated from >1 μg total RNA by random hexamer primers (Subheading 3.3, **step 5**).

2. Forward and reverse primer set (preferably intron spanning) (10 μM stock).

3. dNTP mix (10 mM each).

4. 10× Reaction buffer.

5. Taq polymerase.

6. Molecular biology grade agarose.

7. TRIS-Borate-EDTA (TBE) buffer (1×).

8. Ethidium bromide (EtBr).

9. 100 bp DNA marker.

2.6 RT-qPCR of Skipping Efficiency

1. cDNA generated from >1 μg total RNA by random hexamer primers (Subheading 3.3, **step 5**).

2. Forward and reverse primer set for the gene of interest (one of the primers should span the exon boundary of the skipped product (e.g., the primer should consist of the last 10 nucleotides of exon 1, and the first 10 nucleotides of exon 3 of the gene of interest when skipping exon 2)) (10 μM stock).

3. Forward and reverse primer set for a suitable reference gene (e.g., *GUSB* or *GAPDH*) (10 μM stock).

4. 2× SYBR-green PCR mastermix.

3   Methods

RNA isolated from the nucleofected myotube cultures is used to generate cDNA, which can subsequently be used for RT-PCR or RT-qPCR analysis of exon skipping efficiency. To purify a sufficient amount of total RNA from myotube cultures, we usually transfer cells to six-well plates after nucleofection. Smaller culture vessel might be suitable, or more optimal, for other purposes. We use up to 1×10^6 myoblast cells per 20 μL nucleofection reaction, requiring 1.6×10^7 cells in total for a full 16-well cuvette strip. A T182 culture vessel of high confluence contains about 2×10^7 myoblast cells. A graphical overview of the procedure is outlined in Fig. 1. All steps described in Subheading 3.1, **steps 1–17** should be performed under aseptic conditions. Subheading 3.1, **steps 18** through Subheading 3.2, **step 11** should be performed using suitable personal protection according to local regulations.

3.1 Nucleofection and Maintenance of Cells

This example assumes an experiment involving 1 full 16-well nucleocuvette strip, using 1×10^6 cells per reaction. Scale volumes up or down according to experimental design. Prior to starting the procedure, make sure that the nucleofector buffer has been supplemented as indicated. Prior to handling the cells, make sure that the desired programs to be used have been entered into the 4D-nucleofector core unit to avoid delays in the procedure when handling resuspended cells (*see* **Note 6**).

1. Preferentially subculture myoblasts 1 day prior to nucleofection to ensure proper viability and growth phase of cells. Cells from confluent cultures might exhibit a decreased efficiency or survival.

2. Label wells and add media to the culture vessels used to transfer cells post-nucleofection, e.g., 3 mL proliferation media per well of a 6-well plate. Place plates in a 37 °C CO_2 incubator to warm and equilibrate the media.

3. On the day of the nucleofection experiment, wash the cells with dPBS and trypsinize the cells. Use 2 mL trypsin 0.05% for a

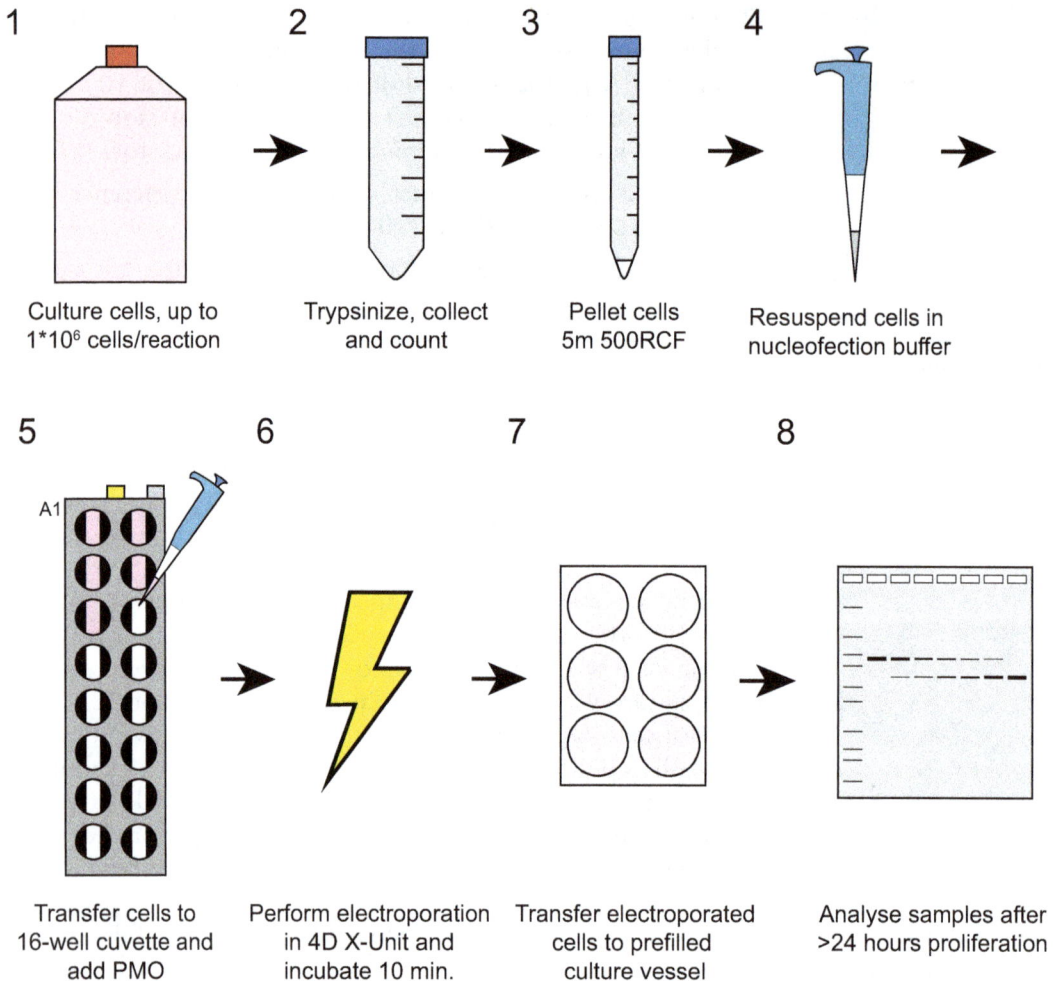

Fig. 1 Schematic workflow of nucleofecting PMO in myoblast cells

T182 and carefully tilt flask to cover all cells. Place cells with trypsin in a 37 °C incubator.

4. When cells are properly detached from the surface (cell line dependent, but generally after 2 min), add resuspension media to a total volume of 10 mL to inhibit trypsin activity.

5. Resuspend cells and transfer to a 50-mL tube (*see* **Note 7**). If multiple flasks are needed to obtain the required number of cells need for the experiment performed, pool cells at this step.

6. Count the number of cells in the 50-mL tube (at least twice) using available cell counting methods. Transfer a total of 1.7×10^7 cells (16 1×10^6 reactions + 1 surplus) to a fresh conical tube, and pellet cells by centrifugation for 5 min at 200 relative centrifugal force (RCF).

7. Aspirate medium completely and gently resuspend cell pellet in 20 µL nucleofection buffer per reaction (i.e., here we use 340 µL for 16 reactions + 1 surplus) (*see* **Note 8**).

8. Aliquot 20 µL cell suspension in each of the chambers in the 16-well nucleofector strip using a P20 micropipette. Avoid creating air bubbles (*see* **Note 9**). When placing the cuvette strip with the yellow notch away from the user, the top left well is well A1 (*see* **Note 10**).

9. Add 1 µL of PMO diluted to the desired concentration. For example, 1 µL PMO from a 1 mM stock will result in a final concentration in the cuvette of 50 µM. Gently mix by stirring with the pipet tip, do **not** vigorously pipet up and down (*see* **Notes 11** and **12**).

10. Replace the lid on the cuvette strip (*see* **Note 13**) and place it in the 4D-Nucleofector X unit tray (yellow notch away from user, toward the machine).

11. Start nucleofection by selecting start on the 4D-nucleofector core unit touch screen.

12. After the machine has finished electroporating, check the status on the display. Green crosses mean the samples have been electroporated without issues. When other symbols are shown, an error might have occurred. Refer to the 4D-nucleofector user manual.

13. Leave the nucleocuvette strip on the bench for 10 min for the cells to recover (*see* **Note 14**).

14. Gently add 180 µL prewarmed media to each of the wells of the cuvette strip.

15. Carefully resuspend (maximum 2–3 times up and down) the cells in the cuvette, and transfer to the culture vessel prepared in Subheading 3.1, **step 2** (*see* **Note 15**).

16. Allow cells to proliferate for 24 up to 72 h and monitor viability of electroporated cells (*see* **Note 16**). Let cells proliferate until reaching 100% confluency.

17. Upon reaching confluency, replace myoblast proliferation media with differentiation media to induce myotube formation.

18. After 72 h, wash cells with dPBS and lyse in 500 µL TRI-reagent (for a 6 well plate).

19. Collect samples using a cell scraper and transfer to 1.5-mL microcentrifuge tubes on ice.

20. Store samples (−80 °C) or immediately isolate total RNA from cells according to available protocols (e.g., 3.2).

3.2　RNA Isolation

1. Thaw the lysate obtained at Subheading 3.2, **step 17** and add 1/5th volume (100 μL) of chloroform (*see* **Notes 17** and **18**).
2. Mix thoroughly by shaking for 30 s.
3. Incubate on ice for 5 min.
4. Centrifuge samples for 15 min at 16,000 RCF at 4 °C.
5. Carefully transfer the upper aqueous phase to a new microcentrifuge tube, without disturbing the organic- and interphase.
6. Add an equal volume of 2-propanol to transferred aqueous phase.
7. Mix well and incubate for >30 min on ice, then centrifuge for 15 min at 16,000 RCF at 4 °C.
8. Discard supernatant and wash pellet with 1 mL 70% EtOH.
9. Centrifuge for 5 min at 16,000 RCF at 4 °C.
10. Discard supernatant and air-dry pellet.
11. Dissolve the pellet in RNase-free water (e.g., 25 μL).
12. Determine RNA concentration using a Nanodrop ND-1000.
13. Store RNA at −80 °C (*see* **Note 19**).

3.3　cDNA Synthesis

1. To generate first-strand cDNA using random hexamer using at least 1 μg of total RNA in a 20 μL reaction. In a 12.5 μL reaction mix the following components (*see* **Note 20**):
   (a) X μL RNA for a total of 1 μg.
   (b) 1 μL dNTP mix (10 mM each).
   (c) 1 μL Random hexamer primers (N6) (20 ng/μL).
   (d) X μL RNase-free water up to 12.5 μL.
2. Incubate reaction for 5 min at 65 °C. Chill on ice for 2 min.
3. To each tube add:
   (a) 4 μL 5× RT-reaction buffer.
   (b) 1 μL Reverse transcriptase enzyme.
   (c) 0.5 μL RNase inhibitor.
4. Incubate at 42 °C for 1 h.
5. Incubate at 85 °C for 5 min.
6. Dilute the 20 μL cDNA reaction to a final volume of 100 μL with ultrapure MQ.
7. Store cDNA at −20 °C.

3.4　RT-PCR Analysis

1. Set up RT-PCR (25 μL reactions) according to the follow set-up per reaction (*see* **Note 21**):
   (a) 2.5 μL 10× Reaction buffer.
   (b) 1 μL Forward primer (10 μM).

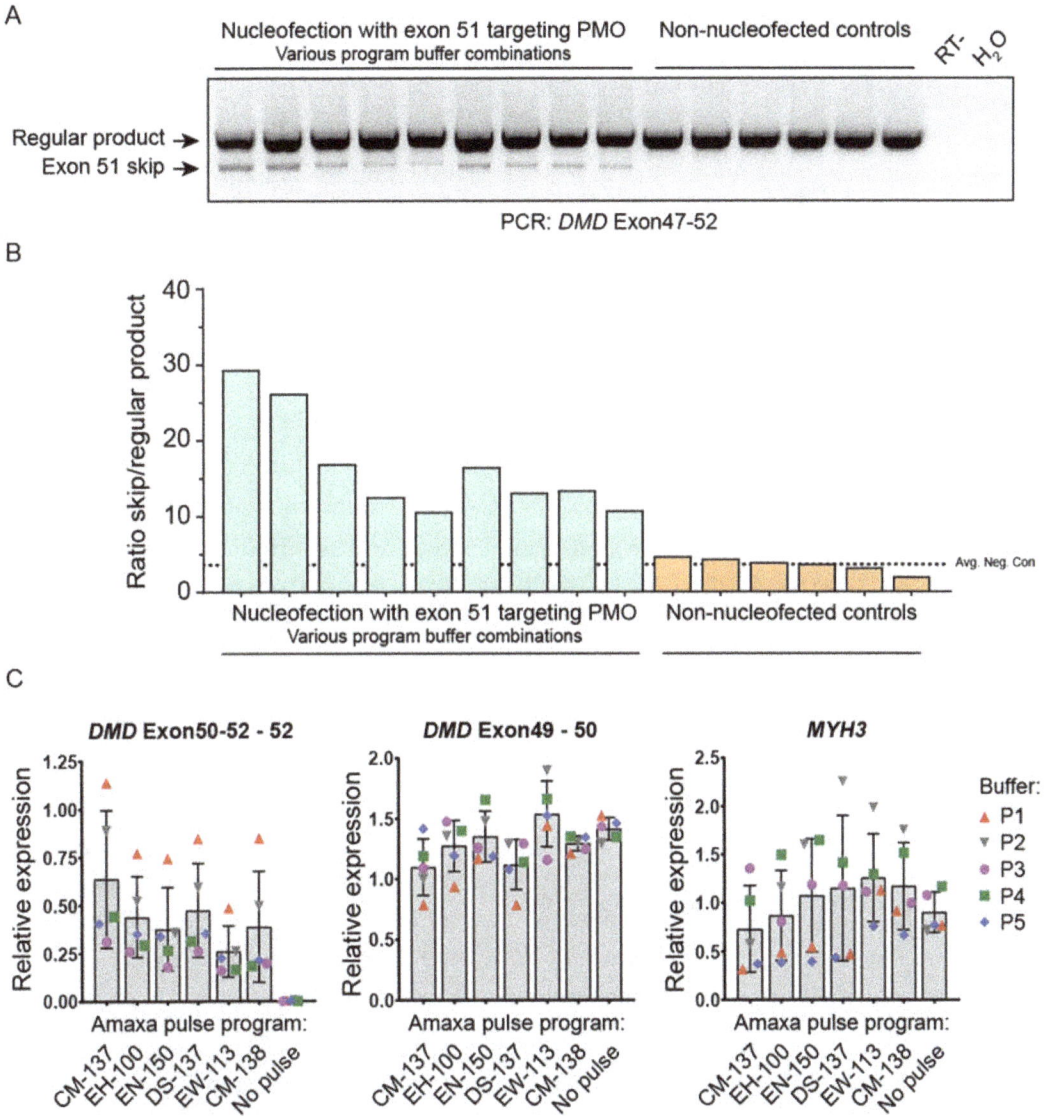

Fig. 2 Example of the result of RT-PCR analysis after successful nucleofection of a *DMD* exon 51 targeting PMO in KM155 myotubes using various programs of the 4D-nucleofector. (**a**) Agarose gel electrophoresis of endpoint RT-PCR products using primers specific for *DMD* exon 47 through 52. Skipping of exon 51 will lead to the production of a smaller PCR product as indicated. Different lanes consist of different experimental combination of nucleofection buffers and programs. (**b**) Approximate quantification of agarose gel shown in A by FIJI analysis. Ratios of Exon 51 skipped product intensities over the regular (exon 51 containing) product are plotted. The average of the negative controls (no PMO/no pulse samples (orange bars)) is shown as a dotted line. (**c**) RT-qPCR analysis of *DMD* exon 51 skipping in KM155 myotubes with 6 nucleofection programs and 5 nucleofection buffers. Exon skipping was determined with a primer set only amplifying the *DMD* transcript without exon 51 (Exon 50-52F + Exon 52R). A primer set specific for *DMD* exon 49 through 50 shows all *DMD* transcript, and *MYH3* is used as a marker for myogenic differentiation. Cells resuspended in nucleofection buffer not subjected to an electric pulse were used as a negative control. In our hands, exon skipping efficiency was highest using a combination of program CM-137 and buffer P1

(c) 1 μL Reverse primer (10 μM).

(d) 1 μL dNTP mix (10 mM each).

(e) 0.2 μL Taq polymerase (5 U/μL).

(f) 10 μL cDNA from Subheading 3.3, **step 5**.

(g) 9.3 μL ultrapure water (up to 25 μL).

2. Gently mix and place samples in a thermal cycler using the following program:

(a) 1: 5 min, 95 °C—initial melt.

(b) 2: 30 s, 95 °C—melt.

(c) 3: 30 s, 60 °C—annealing (change for specific primers).

(d) 4: 40 s, 72 °C—extension (~30 s per kb).

(e) 5: Go to **step 2**, 34 additional times (35 cycles total).

(f) 6: 5 min, 72 °C—final extension.

3. Analyze PCR samples by standard ethidium bromide (EtBr) agarose gel electrophoresis, using a 2% agarose gel (for products <1000 bp) in TBE buffer at 125 V (Fig. 2) (*see* **Note 22**). Alternatively, accurate analysis of signal intensity can be measured by use of, e.g., an Agilent 2100 Bioanalyzer (*see* **Note 23**).

3.5 RT-qPCR Analysis

1. For RT-qPCR analysis, measure the gene of interest and at least one reference gene. Always measure a technical triplicate of each cDNA-primer combination. Set up RT-qPCR according to the following set-up per reaction (we use 8 μL reactions in a 384-well plate):

(a) 4 μL 2× SYB green master mix.

(b) 1 μL Forward primer (10 μM).

(c) 1 μL Reverse primer (10 μM).

(d) 2 μL cDNA from Subheading 3.3, **step 5**.

2. Seal the plate and mix by inversion and briefly spinning down the plate in a centrifuge.

3. Place samples in a real-time PCR enabled thermal cycler and run the following program:

(a) 1: 5 min, 95 °C—initial melt.

(b) 2: 10 s, 95 °C—melt.

(c) 3: 30 s, 60 °C—Annealing and extension (change temperature for specific primers).

(d) 4: Read plate, go to **step 2**, 39 additional times (40 cycles total).

(e) 5: Melt curve analysis.

4. A suitable primer pair for the skipped product will yield little to no signal in the un-nucleofected control samples and will have increased abundance when the exon was successfully skipped.

4   Notes

1. The procedure described herein has been optimized and validated using immortal muscle cell line KM155, which has been described previously [11, 12]. Use of this protocol for primary muscle cells or other immortalized cell lines might require additional optimization.

2. Different media are described in the literature for proliferation and/or differentiation of myoblasts. These different media can have considerable effects on gene expression and morphology. For example, differentiation of myoblasts with 2% FCS leads to slower formation of multinucleated myotubes compared to 2% KOSR. However, myotubes formed with 2% KOSR are harder to handle due to faster release from the vessel surface, presumable due to spontaneous contraction of the myotubes. We recommend that different media are tested to compare optimal experimental conditions.

3. For electroporation of adherent cells, a few of the options available are the Lonza 4D nucleofector Y-unit or the Nepagene NEPA21 electroporator.

4. Delivery of PMOs by nucleofection can be prohibitively expensive for many laboratories, as prices for consumables are steep compared to common laboratory transfection reagents, some of which can be cheaply prepared in-house. On the other hand, if the material to be transfected is expensive or rare, nucleofection allows for relatively high concentrations to be applied directly to the cells due to the low reaction volume (as low as 20 μL), saving costs on oligonucleotides.

5. While small molecules such as siRNAs and AONs are easily delivered into myoblasts by electroporation (e.g., siRNA, PMOs) or transfection (e.g., siRNA, 2OMePS AONs), delivery of (large) plasmids is notoriously inefficient.

6. For our immortalized myoblast cell line (KM155), we have tested various different nucleofector pulse programs in combination with the nucleofection buffers present in the primary cell line optimization kit (i.e., P1, P2, P3, P4, and P5). We noticed that the trend of *DMD* exon skipping efficiency was similar for each buffer and depended largely on the nucleofection program used for the 4D X unit (Fig. 2c, bars). Buffers did however largely contribute to overall efficiency of a set of nucleofections (Fig. 2c, individual data points per bar).

7. When trypsinizing cells from the culture vessel, dissociate cells by controlled but forceful pipetting with a 10-mL pipet and a pipetboy to generate a single-cell suspension. Cell clumps will severely hamper nucleofection.

8. Cells should not be kept in nucleofection buffer for a prolonged amount of time, as this can reduce efficiency and viability. After diluting the cell pellet in nucleofection buffer, working swiftly and accurately is key.

9. At any step of the procedure, but especially upon loading of the nucleocuvette, avoid air bubbles in the reagents containing cells. Cavitation shearing and electric arcing can occur when air bubbles are present. Loading the cuvette with a P20 and reverse pipetting (i.e., not pressing the piston to the second stop) are useful for avoiding air bubbles in the cuvette strip.

10. We have tested the re-use of single cuvette strips for the delivery of the same PMO without any noticeable effect on delivery efficiency or cell viability. If the proprietary nucleofector solutions have not finished upon finishing the cuvette strips of a kit, this method can prolong the use of a single ordered kit. To clean a cuvette strip: after transferring cells from a nucleofection experiment from the 16-well strip to a culture vessel, immediately submerge the strip (without lid) in a-100 mL glass bottle filled with sterile milli-Q (MQ). Shake vigorously for 30 s. Decant MQ and add 70% EtOH for disinfection and shake vigorously. Do not leave the cuvette strip in ethanol for extended periods of time to prevent damage to the electrode. Decant EtOH and rinse twice with sterile MQ, airdry strip and store for future use. We have not tested degradation of efficiency after more than two re-uses. Different nucleofection programs with higher voltages might affect electrode degradation and result in poor efficiency upon re-use. Re-usability should always be tested in the experimental setup used to avoid problems in reproducibility.

11. As an alternative to loading the PMO directly in the cell suspension, it is possible to mix PMO, cells and nucleofection buffer prior to dispensing in the 16-wells nucleocuvette. For example, in bulk when multiple pulse programs are tested, or in sterile strips/plates for testing large amounts of different PMOs. For the latter, it is possible to transfer cells to the nucleocuvette by use of a multichannel pipette.

12. It is possible to add the PMO to the nucleocuvette wells prior to adding the cells, in which addition of the cells and handling of the strip will sufficiently mix the substrate with the cell suspension. However, as PMO solutions tend to be slightly viscous, it can be hard to reliably dispense a small volume (e.g., 1 or 2 µL) in a dry well, compared to dispensing in solution.

13. The cell suspension (20 µL) in the nucleocuvette strip should cover the entire bottom of the well in the nucleocuvette. After loading the cell suspension and prior to placing the cuvette in the X-unit, tap it several times on the working surface with appropriate force.

14. Excessive cell death may warrant changes in recovery steps. For example, after electroporation, add 180 µL warm culture media to each well of the cuvette strip, and incubate the strip 10 min at 37 °C. Afterwards, transfer cells carefully as described above.

15. Cells are fragile immediately after electroporation. Avoid shear stress by over-resuspension with small diameter tip orifices. Use a P200 or P1000 micropipette, resuspend two or three times in the cuvette, and immediately transfer to the pre-warmed culture plate. Avoid creation of air bubbles by aggressive pipetting.

16. Certain cell types might show increased viability after nucleo-fection when treated with ROCK inhibitors to block apoptosis 4 h prior and 24 h post nucleofection.

17. RNA is extremely sensitive to degradation by environmental RNases, such as present on human skin. When handling/preparing RNA samples, always wear suitable gloves, and clean work surfaces with RNase removal agents (e.g., RNaseZAP).

18. To compare samples properly, it is important to process sample sets in the same way. This includes but is not limited to simultaneous isolation of RNA, simultaneous generation of cDNA, identical input of RNA for cDNA reactions, etc.

19. For assessing the efficiency of *DMD* exon skipping with RT-PCR, we generally do not incorporate a DNase I treatment step in the RNA isolation protocol.

20. Generation of cDNA can be performed in PCR strips with caps, allowing for easy incubations in a thermal cycler and transfer of sample to another PCR strip with a multichannel pipette.

21. For analysis of exon skipping efficiency by RT-PCR, we strongly recommend the use of single-reaction RT-PCR instead of nested PCR amplifications. Nested PCR amplifications have a high tendency to induce preferential amplification bias and are less quantitatively reliable [13]. Different polymerases and protocols might be better suited for specific transcripts and should be tested.

22. When using software such as ImageJ/FIJI to analyze agarose gel band intensities, it is of the utmost importance that the user keeps the limitations of the software in mind. Saturated signal on the source image will lead to misleading results if analyzed

incorrectly. Therefore, while useful for estimating relative intensity, many applications will require more sensitive methods to reliably quantify exon skipping efficiency. Furthermore, detection of DNA in agarose gels is facilitated by the amount of EtBr intercalating with the DNA, resulting in signal when exposed to UV light. As shorter PCR products bind less EtBr, their intensity is inherently underestimated when measured in analysis software.

23. Prior to analyzing PCR products on an Agilent 2100 bioanalyzer, it is advised to always run some of the sample on an EtBr agarose gel to confirm successful PCR amplification.

References

1. Aartsma-Rus A, van Ommen GJ (2007) Antisense-mediated exon skipping: a versatile tool with therapeutic and research applications. RNA 13(10):1609–1624. https://doi.org/10.1261/rna.653607

2. Douglas AG, Wood MJ (2013) Splicing therapy for neuromuscular disease. Mol Cell Neurosci 56:169–185. https://doi.org/10.1016/j.mcn.2013.04.005

3. Jirka SM, Heemskerk H, Tanganyika-de Winter CL, Muilwijk D, Pang KH, de Visser PC, Janson A, Karnaoukh TG, Vermue R, t Hoen PA, van Deutekom JC, Aguilera B, Aartsma-Rus A (2014) Peptide conjugation of 2'-O-methyl phosphorothioate antisense oligonucleotides enhances cardiac uptake and exon skipping in mdx mice. Nucleic Acid Ther 24(1):25–36. https://doi.org/10.1089/nat.2013.0448

4. Neil EE, Bisaccia EK (2019) Nusinersen: a novel antisense oligonucleotide for the treatment of spinal muscular atrophy. J Pediatric Pharmacol Ther 24(3):194–203. https://doi.org/10.5863/1551-6776-24.3.194

5. Aartsma-Rus A, Krieg AM (2017) FDA approves Eteplirsen for Duchenne muscular dystrophy: the next chapter in the Eteplirsen saga. Nucleic Acid Ther 27(1):1–3. https://doi.org/10.1089/nat.2016.0657

6. Aartsma-Rus A, Van Deutekom JC, Fokkema IF, Van Ommen GJ, Den Dunnen JT (2006) Entries in the Leiden Duchenne muscular dystrophy mutation database: an overview of mutation types and paradoxical cases that confirm the reading-frame rule. Muscle Nerve 34(2):135–144. https://doi.org/10.1002/mus.20586

7. Bladen CL, Salgado D, Monges S, Foncuberta ME, Kekou K, Kosma K, Dawkins H, Lamont L, Roy AJ, Chamova T, Guergueltcheva V, Chan S, Korngut L, Campbell C, Dai Y, Wang J, Barišić N, Brabec P, Lahdetie J, Walter MC, Schreiber-Katz O, Karcagi V, Garami M, Viswanathan V, Bayat F, Buccella F, Kimura E, Koeks Z, van den Bergen JC, Rodrigues M, Roxburgh R, Lusakowska A, Kostera-Pruszczyk A, Zimowski J, Santos R, Neagu E, Artemieva S, Rasic VM, Vojinovic D, Posada M, Bloetzer C, Jeannet PY, Joncourt F, Díaz-Manera J, Gallardo E, Karaduman AA, Topaloğlu H, El Sherif R, Stringer A, Shatillo AV, Martin AS, Peay HL, Bellgard MI, Kirschner J, Flanigan KM, Straub V, Bushby K, Verschuuren J, Aartsma-Rus A, Béroud C, Lochmüller H (2015) The TREAT-NMD DMD global database: analysis of more than 7,000 Duchenne muscular dystrophy mutations. Hum Mutat 36(4):395–402. https://doi.org/10.1002/humu.22758

8. Tuffery-Giraud S, Béroud C, Leturcq F, Yaou RB, Hamroun D, Michel-Calemard L, Moizard MP, Bernard R, Cossée M, Boisseau P, Blayau M, Creveaux I, Guiochon-Mantel A, de Martinville B, Philippe C, Monnier N, Bieth E, Khau Van Kien P, Desmet FO, Humbertclaude V, Kaplan JC, Chelly J, Claustres M (2009) Genotype-phenotype analysis in 2,405 patients with a dystrophinopathy using the UMD-DMD database: a model of nationwide knowledgebase. Hum Mutat 30(6):934–945. https://doi.org/10.1002/humu.20976

9. Aung-Htut MT, McIntosh CS, West KA, Fletcher S, Wilton SD (2019) In vitro validation of phosphorodiamidate morpholino oligomers. Molecules 24(16):2922. https://doi.org/10.3390/molecules24162922

10. Hori S-I, Yamamoto T, Waki R, Wada S, Wada F, Noda M, Obika S (2015) Ca2+ enrichment in culture medium potentiates

effect of oligonucleotides. Nucleic Acids Res 43(19):e128. https://doi.org/10.1093/nar/gkv626

11. Mamchaoui K, Trollet C, Bigot A, Negroni E, Chaouch S, Wolff A, Kandalla PK, Marie S, Di Santo J, St Guily JL, Muntoni F, Kim J, Philippi S, Spuler S, Levy N, Blumen SC, Voit T, Wright WE, Aamiri A, Butler-Browne G, Mouly V (2011) Immortalized pathological human myoblasts: towards a universal tool for the study of neuromuscular disorders. Skelet Muscle 1:34. https://doi.org/10.1186/2044-5040-1-34

12. Echigoya Y, Lim KRQ, Trieu N, Bao B, Miskew Nichols B, Vila MC, Novak JS, Hara Y, Lee J, Touznik A, Mamchaoui K, Aoki Y, Takeda S, Nagaraju K, Mouly V, Maruyama R, Duddy W, Yokota T (2017) Quantitative antisense screening and optimization for exon 51 skipping in Duchenne muscular dystrophy. Mol Ther 25(11):2561–2572. https://doi.org/10.1016/j.ymthe.2017.07.014

13. Spitali P, Heemskerk H, Vossen RHAM, Ferlini A, den Dunnen JT, t Hoen PAC, Aartsma-Rus A (2010) Accurate quantification of dystrophin mRNA and exon skipping levels in duchenne muscular dystrophy. Lab Investig 90(9):1396–1402. https://doi.org/10.1038/labinvest.2010.98

11

Generation of Protein-Phosphorodiamidate Morpholino Oligomer Conjugates for Efficient Cellular Delivery via Anthrax Protective Antigen

Valentina Palacio-Castañeda, Roland Brock and Wouter P. R. Verdurmen

Abstract

Phosphorodiamidate morpholino oligomers (PMOs) offer great promise as therapeutic agents for translation blocking or splice modulation due to their high stability and affinity for target sequences. However, in spite of their neutral charge as compared to natural oligonucleotides or phosphorothioate analogs, they still show little permeability for cellular membranes, highlighting the need for effective cytosolic delivery strategies. In addition, the implementation of strategies for efficient cellular targeting is highly desirable to minimize side effects and maximize the drug dose at its site of action. Anthrax toxin is a three-protein toxin of which the pore-forming protein anthrax protective antigen (PA) can be redirected to a receptor of choice and lethal factor (LF), one of the two substrate proteins, can be coupled to various cargoes for efficient cytosolic cargo delivery. In this protocol, we describe the steps to produce the proteins and protein conjugates required for cytosolic delivery of PMOs through the cation-selective pore generated by anthrax protective antigen. The method relies on the introduction of a unique cysteine at the C-terminal end of a truncated LF (aa 1–254), high-yield expression of the (truncated) toxin proteins in *E. coli*, functionalization of a PMO with a maleimide group and coupling of the maleimide-functionalized PMO to the unique cysteine on LF by maleimide-thiol conjugation chemistry. Through co-administration of PA with LF-PMO conjugates, an efficient cytosolic delivery of PMOs can be obtained.

Key words Antisense, Anthrax toxin, Protective antigen, Phosphorodiamidate morpholino oligomers, DNA analog, Drug delivery, Cellular internalization, Bioconjugate chemistry

1 Introduction

Phosphorodiamidate morpholino oligomers (PMOs) are uncharged DNA analogs with therapeutic potential due to their ability to specifically bind to target sites on RNA. By steric inhibition of translation initiation complexes, PMOs can block translation. Alternatively, by targeting sites associated with splicing of pre-mRNAs, PMOs can mediate splice modulation and thereby correct the consequences of splicing mutations at the pre-mRNA level, for instance those in inherited retinal dystrophies [1]. PMOs have several qualities that are excellent for therapeutic development, including nuclease-resistance, long-term activity, low toxicity, and high specificity [2, 3]. However, a major challenge remains, which is achieving an efficient cellular delivery, particularly in vivo. PMOs are neutral molecules that because of their size are impermeable to cellular membranes. Delivery approaches that have been developed up to now include scraping of cells, particle-based approaches, and cell-penetrating peptide (CPP)-based delivery [3–5]. Cell scraping cannot be translated to in vivo applications and particle-based approaches suffer from delivery challenges in vivo such as poor tissue penetration and liver enrichment [6]. CPP-mediated delivery has demonstrated potential, but still does not target specific cell-surface receptors, indicating the need for a novel approach.

Recently, several groups have demonstrated that anthrax toxin, a sophisticated protein-based molecular machine that has evolved to efficiently deliver toxic catalytic proteins into the cytosol, can be employed for the functional delivery of various types of cargoes, including antisense oligonucleotides (AON) [7, 8]. The full anthrax toxin consists of three proteins: a pore-forming protein, called protective antigen (PA), that generates cation-selective pores and two enzyme components [9], called lethal factor (LF) and edema factor (EF). LF and EF in turn consist of two domains: the first domain binds the protein pore and the second domain is the enzymatically active domain and is thus responsible for the actual toxicity. For the delivery of PMOs via this mechanism, only PA and the PA pore-binding domain of LF are needed as protein components. For LF, this means that the enzymatic (toxicity-causing) domain of LF is replaced with a PMO.

In this chapter, we describe the preparation of the components needed to mediate cytosolic delivery of PMOs by the anthrax toxin translocation mechanism (Fig. 1a). The individual protein components, PA and LF (1–254) are produced in high quantity in soluble form in the cytosol of *E. coli*. LF (1–254) is by itself cysteine-free, so through the introduction of a unique C-terminal cysteine, a site-specific conjugation via maleimide-thiol conjugation chemistry can be achieved. Maleimide-thiol conjugation chemistry is useful for coupling biologically active molecules because it is fast, highly selective and it can be done in physiological buffers at 37 °C or at 4 °C [10].

To enable the conjugation of the PMO to the protein, PMOs containing a primary amine at the 3′ end are functionalized with a maleimide moiety through a bifunctional linker containing an NHS-ester and a maleimide group separated by a cyclohexane spacer. After coupling maleimide-functionalized PMOs with LF, uncoupled PMO is removed via dialysis, producing LF-PMO conjugates that can be delivered to the cytosol via anthrax protective antigen (Fig. 1b).

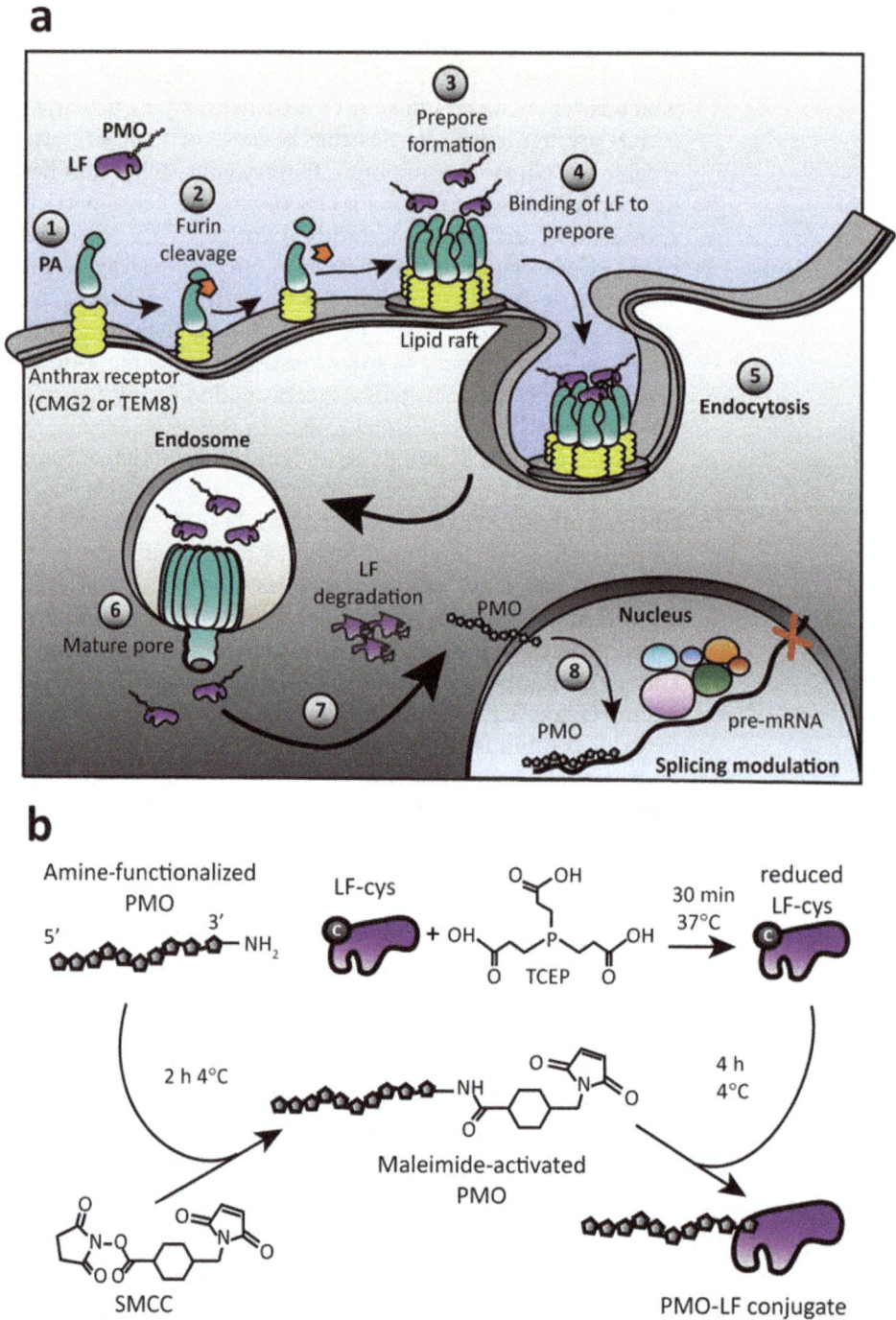

Fig. 1 Cytosolic delivery of PMO using the anthrax toxin mechanism. (**a**) Schematic representation of the cytosolic delivery of the LF-PMO conjugate via anthrax protective antigen. Numbers indicate the distinct steps in the delivery process. (**b**) Schematic representation of the coupling of a PMO to LF-cys. *LF* lethal factor, *PA* protective antigen, *PMO* phosphorodiamidate morpholino oligomer, *SMCC* succinimidyl 4-(*N*-maleimido-methyl)cyclohexane-1-carboxylate, *TCEP* tris(2-carboxyethyl)phosphine

2 Materials

All buffers should be prepared with double-distilled water to ensure highly pure buffers. It is not necessary to work under sterile conditions. However, to perform experiments in the absence of antibiotics, make sure that the final conjugates are filter-sterilized before use.

2.1 Protein Expression

1. BL21(DE3) competent cells (*see* **Note 1**).

2. *E. coli* expression vector coding for anthrax protective antigen (*see* **Note 2**).

3. *E. coli* expression vector coding for LF (1–254) with a C-terminal cysteine (LF-cys) (*see* **Note 3**).

4. LB agar plates: add demi water to 10 g peptone, 5 g yeast, 8 g NaCl, and 15 g agar until a volume of 1 L is reached. Mix well, autoclave and store at room temperature. To make plates, heat the LB agar in the microwave until fully liquid, allow to cool until it is lukewarm before adding the antibiotic of choice. Pour liquid into a Petri dish until you reach a thickness of approximately 0.5 cm. Wait until the agar solidifies and store upside down at 4 °C.

5. 2× YT medium: For 1 L, add demi water to 16 g peptone, 10 g yeast, and 5 g NaCl until a volume of 1 L is reached. Mix well, autoclave, and store at room temperature.

6. Terrific Broth (TB): 12 g/L tryptone, 24 g/L yeast extract, and 4 mL/L glycerol, 17 mM KH_2PO_4 and 72 mM K_2HPO_4. Store at room temperature (*see* **Note 4**).

7. Autoclaved 20% (w/v) glucose solution in demi water.

8. Isopropyl β-D-1-thiogalactopyranoside (IPTG): 1 M stock solution in double-distilled water. Aliquot and store at −20 °C.

9. High-speed centrifuge.

10. Incubator shaker.

2.2 Protein Purification

1. 4-(2-Hydroxyethyl)-1-piperazineethanesulfonic acid (HEPES)-buffered saline (HBS)-wash: 50 mM HEPES, pH 8.0, 150 mM NaCl, 1 mM $MgCl_2$, 20 mM imidazole. Adjust the pH with NaOH to reach pH 8.0. Store at 4 °C. The buffer can be stored at 4 °C for up to 12 months.

2. HBS-resuspension buffer: HBS wash containing 250 μg/mL lysozyme and 1× complete EDTA-free protease inhibitor cocktail. Prepare fresh on the day of the experiment.

3. HBS-low salt: 50 mM HEPES, pH 8.0, 20 mM NaCl, 20 mM imidazole. Adjust the pH with NaOH to reach pH 8.0. Store at 4 °C. The buffer can be stored at 4 °C for up to 12 months.

4. HBS-high salt: 50 mM HEPES, pH 8.0 1 M NaCl, 20 mM imidazole. Adjust the pH with NaOH to reach pH 8.0. Store at 4 °C. The buffer can be stored at 4 °C for up to 12 months.

5. PBS-elution buffer: PBS, pH 7.4, 300 mM imidazole. Adjust the pH with HCl to reach pH 7.4. Store at 4 °C. The buffer can be stored at 4 °C for up to 12 months.

6. Ni-NTA superflow resin (e.g. Qiagen) equilibrated with HBS wash.

7. His-tagged tobacco etch virus (TEV) protease (commercially available).

8. PBS-EDTA-DTT: PBS containing 0.5 mM ethylenediamine-tetraacetic acid (EDTA) and 1 mM dithiothreitol (DTT). Prepare fresh for each experiment.

9. Sonicator.

10. Low-protein binding syringe filter for small volumes (<1 mL) with 0.22 μm pore size.

11. Äkta Pure or alternative chromatography system with UV detector.

12. Superdex 200 10/300 GL column or equivalent column.

13. 4 mL centrifugal filter with a 30 kDa cut-off.

2.3 Protein-PMO Conjugation

1. Tris-buffered saline (TBS): 20 mM Tris-HCl, pH 7.4, 150 mM NaCl. Adjust the pH with NaOH to reach pH 7.4 Store at 4 °C. The buffer can be stored at 4 °C for up to 12 months.

2. 7 kDa MWCO ZEBA spin columns (Thermo Fisher Scientific).

3. Degassed HBS: 50 mM HEPES, pH 7.2, 150 mM NaCl (*see* **Note 5**). Store at 4 °C. The buffer can be stored at 4 °C for up to 12 months. Degas again for 5–10 min before every use.

4. Phosphate-buffered saline (PBS): 10 mM phosphate, pH 7.2, 137 mM NaCl, 2.7 mM KCl. Store at room temperature. Adjust the pH with HCl to set the pH to 7.2 upon dilution from a 10× PBS stock solution prepared with 14.4 g/L Na_2HPO_4, 2.4 g/L KH_2PO_4, 80 g/L NaCl, and 2 g/L KCl, which gives a pH of ~6.8. The buffer can be stored at least for 1 year at room temperature.

5. 1 M lysine solution: For 10 mL, add double-distilled water to 1.46 g lysine until the volume reaches 10 mL. Store at −20 °C. The solution can be kept at −20 °C for at least 1 year.

6. 0.5 M tris(2-carboxyethyl)phosphine (TCEP) solution in double-distilled water, pH 7.0 (*see* **Note 6**) Store aliquots at −20 °C.

7. PMO with a 3′ primary amine modification (Gene tools).

8. Succinimidyl 4-(N-maleimidomethyl)cyclohexane-1-carboxyl-ate (SMCC) linker (Thermo Fisher Scientific).

9. Dialysis membrane (e.g. 20 kDa MWCO).

3 Methods

3.1 Expression of PA and LF-cys

1. Mix 20 μL of BL21(DE3) competent *E. coli* bacteria with 0.5 μL of pure plasmid encoding either PA or LF-cys in a centrifuge tube and place on ice for 15–30 min.

2. Heat shock for 45 s in a water bath at 42 °C, place back on ice for 2 min and subsequently add 200 μL of LB medium. Place horizontally in an incubator shaker at 37 °C for 45 min and plate 10 μL out on an LB agar plate containing the appropriate type and concentration of antibiotic. After an overnight growth, confirm a proper density of colonies and place the plate upside down at 4 °C until starting the next step. It is recommended to start the following step on the same day.

3. Prepare a starter culture by picking a single colony and inoculating 50 mL of 2× YT medium containing 1% (w/v) glucose and antibiotic. Let the bacteria grow overnight at 37 °C with shaking at 150 rpm, orbit diameter 50 mm.

4. Use the starter culture to inoculate 1 L of TB supplemented with 0.8% (w/v) glucose and antibiotic to a starting OD_{600} of 0.1.

5. Allow the bacteria to grow at 37 °C with shaking at 150 rpm, orbit diameter 50 mm, until an OD_{600} of 0.7–0.8 is reached. At this point the expression can be induced by adding 100 μM IPTG. Decrease the temperature to 25 °C (*see* **Note 7**).

6. After 4 h of expression, centrifuge the cells for 10 min at $5000 \times g$ in 500 mL containers. Discard the supernatant and resuspend the pellets in 20 mL of medium. Transfer the resuspended cells to 50 mL conical tubes and centrifuge again for 10 min at $5000 \times g$. Discard the medium and snap-freeze the pellets in liquid nitrogen. Store at −80 °C until purification is commenced.

3.2 Purification of PA and LF-cys

PA and LF are susceptible to denaturation or aggregation when exposed to higher temperatures, so it is recommended to perform all of the subsequent steps at 4 °C.

1. Resuspend bacteria pellets from 1 L expression culture in 20 mL HBS-resuspension buffer.

2. Lyse the bacteria on ice by sonication (60 W with 15 pulses of 10 and 30 s pause between pulses in order to avoid excessive heat or formation of foam (*see* **Note 8**).

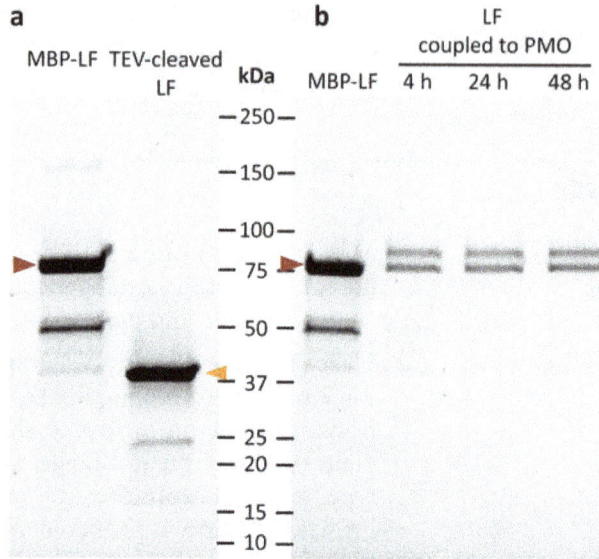

Fig. 2 TEV cleavage of MBP-LF and time dependency of coupling efficiency of LF to PMO. (**a**) SDS-PAGE gel showing LF fused to MBP before TEV cleavage (red arrow) and LF after cleavage and purification via reverse IMAC (orange arrow). (**b**) SDS-PAGE gel illustrating the effect of incubation time on the conjugation efficiency of coupling LF to the PMO. Red arrow indicates 86 kDa band corresponding to MBP-LF before coupling. Green arrow shows a band with increased molecular weight at approximately 92 kDa corresponding to the coupled fraction (MW of PMO: 6.8 kDa). No marked differences are seen between the incubation time and the coupling efficiency, indicating that 4 h incubation is sufficient to achieve 50% coupling. Proteins on gels were visualized by stain-free imaging technology (Bio-Rad). *IMAC* immobilized metal ion affinity chromatography, *LF* lethal factor, *MBP* maltose-binding protein, *PMO* phosphorodiamidate morpholino oligomer, *TEV* tobacco etch virus

3. Centrifuge the bacterial lysates at $28{,}000 \times g$ for 40 min and filter the supernatant using 0.22 μm pore size filters.

4. Incubate the supernatants containing the His-tagged proteins with 1 mL of HBS wash-equilibrated Ni-NTA superflow resin for 1 h on a roller shaker.

5. Transfer the resin-containing solution to a column containing a bottom filter and wash with 20 column volumes (CV) of HBS-low salt, followed by 20 CV of HBS-high salt and 20 CV of HBS wash.

6. Pre-elute the proteins with 1 CV of PBS-elution buffer, followed by eluting with 4×1 CV. Each elution step should be for 10 min on a roller shaker (*see* **Note 9**).

7. Dialyze the eluted proteins overnight against 2 L 1× PBS. In the morning, exchange for fresh PBS and dialyze for two more hours.

8. The purified proteins His$_6$-MBP-LF-cys and His$_6$-MBP-PA can be cleaved with TEV protease to generate LF-cys and PA, respectively (Fig. 2a). To cleave off the His$_6$-MBP add a His-tagged TEV protease at a 1:20 (w/w) ratio in PBS-EDTA-DTT.

9. In order to separate the His$_6$-MBP and the His-tagged TEV from LF-cys or PA, a reverse immobilized metal ion affinity chromatography (IMAC) purification step must be done in which the *unbound* fraction is kept. Because the Ni-NTA resin is incompatible with high concentrations of EDTA and DTT, solutions must first be dialyzed against a 100-fold excess of HBS wash for at least 2 h to sufficiently remove the EDTA and DTT.

10. Incubate the dialyzed samples with 1 mL equilibrated Ni-NTA superflow resin for 1 h on a roller shaker.

11. Collect the unbound fraction, which should contain only LF-cys or PA. An SDS-PAGE gel can be run at this stage in order to confirm correct cleavage (Fig. 2a). To prepare LF-cys for reduction, perform a dialysis against HBS (*see* **Note 10**). Dialyze PA against PBS, after which it will be ready for further purification via size-exclusion chromatography (SEC) (Subheading 3.3).

12. LF needs to be reduced in order to generate the free thiol groups required for maleimide-thiol conjugation chemistry. To reduce LF-cys add a 100-fold molar excess of TCEP, incubate for 30 min at 37 °C, and remove excess TCEP using a 7 kDa MCWO ZEBA spin column. To calculate the amount of TCEP needed, determine the protein concentration using the absorption at 280 nm and the extinction coefficient ε280 of the protein (i.e. 91,680 M^{-1} cm^{-1} for MBP-containing LF and 23,840 M^{-1} cm^{-1} for cleaved LF).

3.3 Purification of PA by Size-Exclusion Chromatography (SEC)

After purification of PA via IMAC, we typically see some co-purification of degraded fragments, which can be removed via SEC (*see* **Note 11**). We generally purify our PA or PA fusion constructs via SEC using an Äkta pure chromatography system equipped with an S200 column (*see* **Note 12**), which yields pure and active proteins [11, 12].

1. Concentrate the sample to a volume of less than 500 μL, which is the maximal volume that can be loaded on an S200 column. Make sure that the protein solution is sufficiently concentrated so that in the maximally 500 μL several milligrams of proteins can be loaded (ideally 2–5 mg). For lower amounts, the separation of individual peaks can become problematic. Filtration of the protein using a low protein-binding syringe filter (0.22 μm pore size) before loading is recommended since protein

aggregates may clog the column. Alternatively, centrifugation of the sample at high speed ($\sim$20,000 $\times$ g) for 5 min and loading the supernatant can be done.

2. Connect the column to the chromatography system drop to drop (i.e. without introducing air) and equilibrate the column with PBS. Connect one or multiple injection loops (i.e. depending on the system) that can hold twice the volume of the sample to the chromatography system. For example, for a 500 μL sample, use a 1 mL injection loop. Wash the loop extensively with PBS before loading the sample.

3. When the sample has been loaded, start the run. A typical speed is 0.5 mL/min. Start fraction collection after the dead volume ($\sim$7–8 mL on an S200 column connected to an Äkta Pure system). A suitable fraction size is 0.5 mL. PA should elute at around 12 mL, which should give a major peak detectable by the UV detector.

4. Run an SDS-PAGE gel with collected fractions of any major peaks to determine the molecular weight and purity of the collected proteins. Full-length PA will have a size of approximately 83 kDa.

5. Combine all fractions that contain pure PA, concentrate with a 4 mL centrifugal filter (cut-off: 30 kDa) until a concentration of 2–5 mg/mL is obtained and freeze in aliquots. Aliquots should be snap-frozen in liquid nitrogen and kept at −80 °C. Thawed aliquots of PA can refrozen in liquid nitrogen at least three times without a noticeable decrease in activity.

3.4 Functionalizing PMO with a Maleimide Moiety

1. Dissolve PMO containing a primary amine in double-distilled water to a concentration of 1 mM. The PMO can be stored either at room temperature or aliquoted and stored at −80 °C for long-term storage (see Note 13).

2. Calculate the amount needed of the SMCC linker for the PMO coupling, taking into account that a 20-fold molar excess of the linker is needed for the reaction. Weigh the desired amount of the SMCC linker and dissolve in anhydrous DMF (e.g. 1.5 mg SMCC in 100 μL of DMF to make a stock solution of 45 mM). Freeze aliquots at −20 °C and only thaw briefly for use. The NHS moiety of the linker is very reactive and may hydrolyze already through the presence of trace amounts of water.

3. Mix the PMO with a 20-fold molar excess of the SMCC linker with minimal dilution (e.g. 900 μM PMO and 18 mM SMCC linker) and incubate for 2 h at 4 °C (see Note 14).

4. Quench unreacted linker by adding lysine to a final concentration of 100 mM using a 1 M lysine solution.

5. Separate the reacted PMO from the unreacted linker using a buffer-exchange procedure (e.g. 7 kDa ZEBA spin column, depending on the molecular weight of the PMO).

3.5 Coupling of Anthrax Lethal Factor to Maleimide-Functionalized PMO

1. Assuming full recovery of the PMO after the ZEBA spin column and equal volume, add a tenfold molar excess of the maleimide-functionalized PMO to the reduced LF-cys proteins (e.g. 800 μM PMO to 80 μM LF) and incubate for 4 h at 4 °C (Fig. 2b) (*see* **Note 15**).

2. Dialyze overnight to remove the unreacted PMO. Exchange dialysis buffer the next morning and dialyze for two more hours. Upon complete removal of the excess uncoupled PMO, the concentration of the PMO-conjugate can be calculated using the absorption coefficient of the PMO at 265 nm, while correcting for the absorption of the protein at 265 nm (*see* **Note 16**). The conjugates can be snap-frozen in liquid nitrogen and stored at −80 °C.

4 Notes

1. There are many *E. coli* strains suitable for high-level protein expression of the proteins described in this chapter, but we have good experience with the strain BL21(DE3). For the proteins in question, BL21(DE3) achieves high-level *and* soluble protein expression upon IPTG-mediated induction of T7 polymerase from a *lacUV5* promoter.

2. We typically express full-length PA fused to an N-terminal His_6-maltose-binding protein (MBP) sequence in a pQIq backbone [11]. MBP serves in this context as a solubility enhancer, leading to tens of mg of soluble protein produced in the cytosol of *E. coli* per liter expression culture in shake flasks. We include a TEV protease site between MBP and protective antigen that can be used to cleave off His_6-MBP and generate native PA.

3. Similar to the expression of PA, we express truncated LF (aa 1–254) fused to an N-terminal His_6-MBP sequence. For maleimide-thiol conjugations, we introduce a single cysteine at the most C-terminal position (LF-cys). Similar to PA, this construct gives us very high levels of soluble expression in the cytosol of *E. coli*.

4. 1 L Terrific broth is prepared by autoclaving 12 g tryptone, 24 g yeast extract, and 4 mL glycerol in 800 mL distilled water, followed by the addition of 100 mL of a separately autoclaved (or filter sterilized) solution of 0.17 M KH_2PO_4 and 0.72 M

K_2HPO_4. Adjust volume to 1 L. Mixing the separately autoclaved solutions can be done at the day of the experiment.

5. Degassing should be done at least for several minutes, which generally gives us good results. We flush with nitrogen gas at a speed which generates some, but not excessive, bubbling. Extending the degassing time to 1 h can be considered for optimal results.

6. Dissolving TCEP hydrochloride at 0.5 M in double-distilled water will result in an acidic solution with a pH between 2 and 3. Adjust the pH of the 0.5 M TCEP solution with concentrated NaOH or KOH to pH 7.0.

7. For expression of PA and LF constructs, we decrease the temperature to 25 °C in order to express maximal amounts of soluble proteins. The growth is much slower at 25 °C than at 37 °C, so it is normal that during 4 h of expression, bacteria will not grow much denser (final OD_{600} between 2 and 3). Increasing the shaking speed improves aeration and can give better yields.

8. Even though the protocol describes lysing of cells by sonication, we typically obtain equally good results when we use French press bacterial cell lysis. With both methods, it is important to avoid excessive heat that may quickly denature or aggregate the proteins.

9. During the preelution, very little His-tagged protein will be eluted since the purpose is to equilibrate the resin with the elution buffer. Nevertheless, we prefer not to discard the preelution fractions since they still may contain *some* protein. Upon measuring the concentration, we make the decision to discard it and/or take it along for dialysis (our typical threshold is an A280 of 0.4).

10. Since the reduction in this protocol is done with TCEP, we prefer to use HBS over PBS because TCEP is more stable in HBS.

11. Next to SEC, PA can also be further purified via ion-exchange chromatography or both sequentially if a very high purity is needed.

12. For purification by SEC, we prefer to use a Superdex 200 10/300 GL column (GE Healthcare), which is suitable for separating proteins ranging in molecular weight from 10 to 600 kDa. We get a good separation when separating PA from its degraded fragments with this column. We have not compared with other comparable columns, which may perform equally well. Similarly, while we work with an Äkta pure system, alternative systems for SEC are likely to work equally well and the protocol described would only require small modifications.

13. PMOs can be dissolved in double-distilled water by pipetting up and down. The solution can be frozen at $-80\,°C$ or kept at room temperature. We have not seen any differences in coupling efficiency following freezing and rethawing.

14. The reaction can be performed at room temperature for 30 min or at 4 °C for 2 h. We get slightly better results when performing the latter protocol.

15. We observed an already high efficiency of coupling after 4 h (Fig. 2b), which did not noticeably increase during longer incubation times. It is possible that longer reaction times may still be better for some protein-PMO conjugations, in which case reaction times up to 48 h can be attempted. Higher excesses of PMO lead to greater efficiency but can be cost limiting. Furthermore, removal of much greater excesses of PMO using dialysis will take significantly longer. Additionally, more concentrated solutions of proteins (up to 5 mg/mL) and PMO are desired, taking into account the solubility of each.

16. PMOs absorb much more strongly than proteins at 265 and 280 nm, implying that absorption measurements at either wavelength will be dominated by the absorption of the PMO. For an accurate estimation of the protein-PMO conjugate, it is therefore important to completely remove the uncoupled PMO, which can be achieved either by extensive dialysis or by ion-exchange chromatography. By measuring the absorption of the conjugate at 265 nm and correcting for the absorption of the protein at 265 nm (for this, measure the pure protein and estimate the $\varepsilon265$ nm as follows: $[(A265\,nm/A280\,nm)*\varepsilon280]$), one can use the extinction coefficient of the PMO at 265 nm (provided by the manufacturer) to calculate the concentration of the protein-PMO conjugate. For a proper correction, take into account the labeling efficiency of the protein through analysis of conjugates by SDS-PAGE.

Acknowledgments

This work was supported by funding from Landelijke Stichting voor Blinden en Slechtzienden, grant number: UitZicht 2017-14.

References

1. Gerard X, Garanto A, Rozet JM et al (2016) Antisense oligonucleotide therapy for inherited retinal dystrophies. Adv Exp Med Biol 854: 517–524

2. Summerton J, Weller D (1997) Morpholino antisense oligomers: design, preparation, and properties. Antisense Nucleic Acid Drug Dev 7(3):187–195

3. Shabanpoor F, McClorey G, Saleh AF et al (2015) Bi-specific splice-switching PMO oligonucleotides conjugated via a single peptide active in a mouse model of Duchenne muscular dystrophy. Nucleic Acids Res 43(1):29–39

4. Partridge M, Vincent A, Matthews P et al (1996) A simple method for delivering morpholino antisense oligos into the cytoplasm of cells. Antisense Nucleic Acid Drug Dev 6(3): 169–175

5. Summerton JE (2005) A novel reagent for safe, effective delivery of substances into cells. Ann N Y Acad Sci 1058:62–75

6. Juliano RL (2016) The delivery of therapeutic oligonucleotides. Nucleic Acids Res 44(14): 6518–6548

7. Dyer PDR, Shepherd TR, Gollings AS et al (2015) Disarmed anthrax toxin delivers antisense oligonucleotides and siRNA with high efficiency and low toxicity. J Control Release 220(Pt A):316–328

8. Wright DG, Zhang Y, Murphy JR (2008) Effective delivery of antisense peptide nucleic acid oligomers into cells by anthrax protective antigen. Biochem Biophys Res Commun 376(1):200–205

9. Young JAT, Collier RJ (2007) Anthrax toxin: receptor binding, internalization, pore formation, and translocation. Annu Rev Biochem 76: 243–265

10. Hansen RE, Winther JR (2009) An introduction to methods for analyzing thiols and disulfides: reactions, reagents, and practical considerations. Anal Biochem 394(2): 147–158

11. Verdurmen WP, Luginbühl M, Honegger A et al (2015) Efficient cell-specific uptake of binding proteins into the cytoplasm through engineered modular transport systems. J Control Release 200:13–22

12. Verdurmen WPR, Mazlami M, Plückthun A (2017) A quantitative comparison of cytosolic delivery via different protein uptake systems. Sci Rep 7(1):13194

12

Design of Bifunctional Antisense Oligonucleotides for Exon Inclusion

Haiyan Zhou ⓘ

Abstract

Bifunctional antisense oligonucleotide (AON) is a specially designed AON to regulate pre-messenger RNA (pre-mRNA) splicing of a target gene. It is composed of two domains. The antisense domain contains sequences complementary to the target gene. The tail domain includes RNA sequences that recruit RNA binding proteins which may act positively or negatively in pre-mRNA splicing. This approach can be designed as targeted oligonucleotide enhancers of splicing, named TOES, for exon inclusion; or as targeted oligonucleotide silencers of splicing, named TOSS, for exon skipping. Here, we provide detailed methods for the design of TOES for exon inclusion, using *SMN2* exon 7 splicing as an example. A number of annealing sites and the tail sequences previously published are listed. We also present methodology of assessing the effects of TOES on exon inclusion in fibroblasts cultured from a SMA patient. The effects of TOES on *SMN2* exon 7 splicing were validated at RNA level by PCR and quantitative real-time PCR, and at protein level by western blotting.

Key words Antisense oligonucleotide, Bifunctional antisense, Pre-mRNA splicing, TOES, Splice switching, Exon inclusion, Exon skipping

1 Introduction

Harnessing antisense oligonucleotides (AONs) to redirect the altered pre-messenger RNA (pre-mRNA) splicing and modulate target gene expression is an efficient therapeutic strategy for genetic disorders associated with alternative splicing. A number of AON approaches have been investigated on redirecting pre-mRNA splicing. The original strategy is to use AONs complementary to a cryptic splice site to prevent its use and favored selection of the authentic site [1]. This approach has been used regularly to alter the proportion of splice isoforms produced from mutated genes or alternative splicing units. In addition to blocking the splice sites, alternative splicing events are often controlled by regulatory proteins bound to exonic and intronic elements located beyond the alternative splice sites. A valid approach is to use AONs to directly target exonic or intronic elements by blocking the binding of regulatory proteins to these elements that are involved in pre-mRNA splicing. This strategy has been successfully used to augment the exon 7 inclusion in *SMN2* gene by using a short AON to target an intronic splicing silencer (ISS) within the gene [2–5]. Nusinersen, an 18-mer AON annealing to the ISS-N1 element in *SMN2* intron 7 is the first antisense drug approved by the US Food and Drug Administration (FDA) for treatment of any types of spinal muscular atrophy (SMA) [6–8]. This strategy has also been proved to be very effective in Duchene muscular dystrophy (DMD) by promoting the skipping of an exon in the *DMD* gene to restore the interrupted reading frame hence partial rescue of the functional dystrophin protein [9, 10]. Three AON drugs, eteplirsen for exon 51 skipping and golodirsen and viltolarsen for exon 53 skipping in the *DMD* gene, have been approved by the FDA for treatment of DMD [11–13].

The other splice switching approach is the use of bifunctional oligonucleotides to increase the number of positively or negatively acting signals in an exon or intron and to regulate the alternative splicing. The oligonucleotides were designed with one domain (the antisense domain) annealing to the target exon or intron, and another domain (the tail domain) containing a sequence that either recruits RNA binding proteins involved in pre-mRNA splicing [14] or is made of a synthetic protein domain covalently linked to the antisense domain [15]. This approach may be designed as targeted oligonucleotide enhancers of splicing (TOES) for exon inclusion [14, 16], or as targeted oligonucleotide silencers of splicing (TOSS) for exon skipping [17].

The effectiveness of TOES as a potential therapy for SMA by augmenting exon 7 splicing in *SMN2* gene has been approved both in vitro in cellular model [14, 16] and in vivo in mouse model [18–20]. A bifunctional oligonucleotide targeted to *SMN2* exon 7 was expressed in transgenic mice within a modified U7 snRNA gene. Expression of the TOES-U7 RNA in a mouse model of SMA produced a substantial improvement in function and lifespan [20]. Two other bifunctional oligonucleotides targeting the intronic splicing silencers in *SMN2* intron 6 and intron 7, which have the dual effects of blocking the silencer and recruiting activator proteins, also showed the potential therapeutic effects in the transgenic mouse models of SMA [18, 19].

We describe here the details in design of bifunctional oligonucleotide for exon inclusion by correcting *SMN2* exon 7 splicing as an example (Fig. 1). TOES oligonucleotides are designed to contain two domains, an antisense domain complementary to sequences of *SMN2* gene and a tail domain comprising sequences known as binding moieties for splicing activator proteins. The following design principles for TOES oligonucleotides are followed: (1) the antisense sequence may anneal to the potential

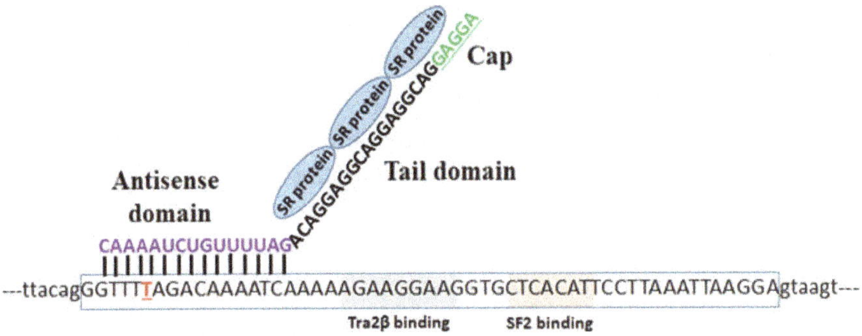

Fig. 1 Design of TOES to promote *SMN2* exon 7 inclusion. The sequence of *SMN2* exon 7 is in upper case and the flanking introns in lower case. Nucleotide 6 in *SMN2* exon 7 is T (in red). Two shaded sequences are the binding sites of Tra2β and SF2/ASF, respectively. TOES is designed with two functional parts, the antisense domain to anneal to nucleotides 2–16 in *SMN2* exon 7, the tail domain containing 3 repeats of "GGAGGAC" motifs to recruit the SR protein SRSF1. Cap contains five nucleotides at the 5′-end of the tail (in green, which is chemically modified)

splicing silencer binding sites in either intron 6, exon 7 or intron 7, and should avoid any splicing enhancer binding sites; (2) a number of splicing enhancer motifs (e.g. SF2/ASF, SRSF1, and hTra2β1) may be included in the tail domain to improve the effectiveness of the oligonucleotides; (3) chemical modification can be applied to the antisense sequence, but not to the tail domain, which may inactive protein binding to the tail domain. The effects on exon inclusion are evaluated at RNA and protein levels in fibroblasts cultured from a patient with type II SMA carrying three copies of *SMN2* gene.

2 Materials

2.1 AON Design

1. Online software to identify splicing motifs, e.g., Human Splicing Finder (http://www.umd.be/HSF/HSF.shtml).

2. Online software to predict the secondary structures of the target gene and AONs (http://rna.urmc.rochester.edu/RNAstructure.html).

3. Online software to calculate oligonucleotide properties on annealing temperature, GC content, and self-complementary (http://biotools.nubic.northwestern.edu/OligoCalc.html).

2.2 Synthesis and Preparation of Bifunctional Oligonucleotides

1. Oligonucleotides are synthesized commercially by Eurogentec Ltd. (www.eurogentec.com).

2. RNase and DNase-free distilled water (*see* **Note 1**).

3. RNase and DNasefree 1.5 mL Eppendorf tubes.

2.3 Culture of Skin Fibroblasts from SMA Patient

1. Growth medium: Dulbecco's modified eagle medium (DMEM), 10% Fetal Bovine Serum (FBS), 1% Glutamax.
2. Trypsin-EDTA.
3. Phosphate-buffered saline (PBS).
4. Incubator set at 37 °C and 5% CO_2.

2.4 Fibroblast Transfection

1. Transfection reagent (e.g. Lipofectamine 2000).
2. Reduced serum medium for transfection (e.g. Opti-MEM).
3. 6 well plate or 35 mm diameter culture dish.
4. Sterile 1.5 mL Eppendorf tubes.

2.5 RNA Extraction

1. RNA isolation kit.
2. β-mercaptoethanol.
3. 70% ethanol (molecular grade).
4. RNase-free water.
5. 1.5 mL RNase-free Eppendorf tubes.
6. NanoDrop spectrophotometer.

2.6 cDNA Synthesis

1. cDNA synthesis kit.
2. Thermocycler.
3. 0.2 mL PCR tubes.

2.7 Polymerase Chain Reaction (PCR)

1. cDNA template from Subheading 2.6.
2. Taq Polymerase.
3. Primers (10 μM forward primer and 10 μM reverse primer). Primers sequences are shown in Table 1.
4. PCR buffer: 10× PCR buffer, 10 mM dNTPs, 50 mM $MgCl_2$, Taq DNA polymerase (5 U/μL), and nuclease-free water.
5. 0.2 mL PCR tubes.
6. Thermocycler.
7. Tris–Borate–EDTA 1× (TBE) buffer.
8. Agarose.
9. DNA gel stain.
10. Loading buffer.
11. DNA ladder.
12. Gel imaging system.

2.8 Quantitative Real-Time PCR

1. cDNA template from Subheading 2.6.
2. qPCR primers (10 pmol/μL each). Sequences are shown in Table 1.

Table 1
Sequences of primers used for exon 7 inclusion quantification by PCR and quantitative real-time PCR

Assay	Products	Sequences (5′–3′)	Annealing Tm (°C)
PCR	Full-length SMN2 (505 bp)	F: CTC CCA TAT GTC CAG ATT CTC TT	55
	Δ7 SMN2 (451 bp)	R: CTA CAA CAC CCT TCT CAC AG	
qRT-PCR	Full-length SMN2 (133 bp)	F: ATA CTG GCT ATT ATA TGG GTT TT	60
	Δ7 SMN2 (125 bp)	R: TCC AGA TCT GTC TGA TCG TTT C	
		F: TGG ACC ACC AAT AAT TCC CC	
		R: ATG CCA GCA TTT CCA TAT AAT AGC C	

3. Universal SYBR Green Master Mix.

4. 96-well real-time PCR plate.

5. Sealing film.

6. Real-Time PCR Thermal Cycler.

2.9 Western Blotting

1. Protein extraction buffer: 0.25% SDS, 75 mM Tris–HCl (pH 6.8), or RIPA buffer.

2. Protease inhibitor cocktail tablets.

3. Pierce BCA Protein Assay Kit.

4. PBST washing buffer (PBS, pH 7.4, 0.1% Tween 20).

5. Mini gel tank and blot transfer set.

6. NuPAGE 10% Bis-Tris precast gels, LDS sample buffer (4×), SDS running buffer (20×), antioxidant, sample reducing buffer, transfer buffer (20×), methanol.

7. Protein molecular weight ladder.

8. PVDF membrane.

9. Odyssey blocking buffer for PVDF membrane blocking.

10. Antibodies: mouse anti-SMN monoclonal antibody (BD Transduction Laboratories), mouse anti-β-tubulin monoclonal antibody (Sigma), IRDye 800CW-conjugated goat anti-mouse secondary antibody (Li-Cor).

11. Odyssey imaging instrument to quantify western blot signals.

3 Methods

3.1 Design of Bifunctional Oligonucleotides

1. Predict the potential binding motifs of the negative splicing regulator heterogeneous nuclear ribonucleoprotein A1 (hnRNP A1) in the target intron or exon sequences, using *Human Splicing Finder* online software.

2. Other splicing repressors, such as intronic splicing silencers and exonic splicing silencers, may also be identified in the literature. A number of annealing sites in intron 6, exon 7, and intron 7 have been reported to augment *SMN2* exon 7 splicing by bifunctional AONs (Table 2) (*see* **Note 2**).

3. AONs, of 15–20 mer in length, are designed to anneal to the potential binding sites of hnRNP A1 or other splicing silencers.

4. The GC content of each AON sequence should be 40–65%, with an ideal content of approximately 60%.

5. Avoid four consecutive "G," strong secondary structure or self-complementary sequences, and self-dimers.

6. Chemical modifications, e.g., 2′-*O*-methyl and locked nucleic acid (LNA), may be applied to the antisense sequence to improve stability and increase binding affinity.

7. Select the tail domain. Examples are listed in Table 2.

8. No chemical modifications are recommended in the tail domain except the cap sequence (Fig. 1) [16] (*see* **Note 3**).

3.2 Transfection of SMA Fibroblasts

1. Seed the cells in a 6-well plate at a concentration of 2×10^5 cells per well, which gives 80% confluence on the next day.

2. Cells are cultured in 2 mL of growth medium for 24 h.

3. 24 h later, change the growth medium to 1 mL Opti-MEM and leave the cells in the incubator during the preparation of transfection mixes.

4. Prepare the transfection reagent mixes in sterile 1.5 mL tubes. For each sample, prepare two mixes: the first mix (Mix A) contains 100 μL Opti-MEM and 1 μL AON at desired concentration (e.g. 1 μL AON at 100 μM to get a 100 nM final concentration). While for the mock control add only 100 μL Opti-MEM. The second mix (Mix B) contains 100 μL Opti-MEM and 5 μL Lipofectamine 2000.

5. Mix the AON-containing tube (Mix A) with the lipofectamine-containing tube (Mix B) at a ratio of 1:1 (100 μL + 100 μL).

6. Incubate the transfection mix for 20 min at room temperature (RT).

7. Add 800 μL Opti-MEM in the transfection mix to top it up to a final volume of 1 mL.

8. Remove Opti-MEM from the 6 well plate and replace with 1 mL transfection mix in each well.

9. Incubate the plate for at least 6 h at 37 °C with 5% CO_2 (*see* **Note 4**).

Table 2
The reported bi-functional AONs designed for SMN2 exon 7 inclusion

Intron/ exon	Antisense binding domain	Sequence of the antisense domain (5′–3′)	Sequence of the tail domain (5′–3′)	Recruiting protein	Reference
Exon 7	hnRNP A1 (ESS)	GAUUUUGUCUAAAAC	ACAGGAGGCAGGAGGCAGGAGGA	SRSF1	[5, 7]
Intron 7	hnRNP A1 binding site (ISS-N1)	GAUUCACUUUCAUAAAUGCUGG	CACACGACACGACACGA	SF2/ASF	[10]
Intron 7	hnRNP A1 binding site (ISS-N1)	GAUUCACUUUCAUAAAUGCUGG	GAAGGAGGGAAGGAGGGAAGGAGG	hTra2β1	[10]
Intron 6	Element 1 (E1)	CUAUAUAGAUAGUUA UUCAACAAA	CACACGACACGACACACGA	SF2/ASF	[9]
Intron 6	Element 1 (E1)	CUAUAUAGAUAGUUA UUCAACAAA	GAAGGAGGGAAGGAGGGAAGGAGG	hTra2β1	[9]
Intron 7	hnRNP A1 binding site (ISS-N1)	AGUAAGAUUCACUUU	UGUGUGUGUGUGUGUGUG	TDP-43	[8]

3.3 Splicing Assay of Bifunctional AONs on SMN2 Exon 7 Inclusion at RNA Level

1. Extract RNA from SMA fibroblasts using RNeasy Mini Kit according to manufacturer's instruction.

2. Reverse transcription: the cDNA is synthesized from 500 ng RNA using cDNA Synthesis kit according to manufacturer's instruction.

3. PCR of *SMN2* transcripts: Use 1 μL cDNA in a 25 μL PCR reaction with 500 pmol of each primer (Table 1), 200 μM of dNTPs, 1.5 mM MgCl$_2$, 2.5 units of Taq polymerase and 1× PCR buffer. The PCR amplification program is as follows: 1 cycle with 3 min at 94 °C (initial denaturation), 25–30 subsequent cycles of 30 s at 94 °C (denaturation), 30 s at 55 °C (annealing), and 30 s at 72 °C (extension), followed by a final 10-min extension at 72 °C. Check an aliquot of the PCR product (5–10 μL) in 1.5% agarose gel electrophoresis and SYBR safe DNA stain using an UV transilluminator. The top band is the full-length *SMN2* product (505 bp). The lower band is the product without exon 7 (Δ7 *SMN2*, 451 bp).

4. Quantitative real-time PCR of full-length and Δ7 *SMN2* transcripts: product specific primers (Table 1), cDNA and 1× PCR Master mix are mixed in a 20 μL PCR reaction. The program includes activation at 95 °C for 3 min, 40 cycles of 95 °C for 10 s, and 60 °C for 1 min. The cycle at which the amount of fluorescence is above the threshold (Ct) is detected. For quantification, it is possible to use the standard curve method produced from serial dilutions of cDNA from untreated SMA fibroblasts, or the ΔΔCt method. Normalize the ratios of full-length *SMN2* and Δ7 *SMN2* to a housekeeping gene (e.g. *HPRT1* or *GAPDH*) (*see* **Note 5**).

3.4 Bifunctional AONs on Restoring SMN Protein Measured by Western Blotting

1. Remove culture medium from the well. Add 100 μL ice-cold lysis buffer to the cells. Keep on ice for 5–10 min. Collect lysates using cell scrapers to fresh 1.5 mL Eppendorf tube and homogenize thoroughly with pipette.

2. Centrifuge at 12,000 × *g* and 4 °C for 10 min. Transfer the supernatant to a fresh tube.

3. Measure protein concentration by a NanoDrop spectrophotometer using the Pierce BCA Protein Assay Kit according to the manufacturer's instructions.

4. Load 5 mg total protein into NuPAGE precast gels and then electrophorese.

5. Transfer electrophoretically separated proteins from the gel to a PVDF membrane.

6. Block the PVDF membrane for 1 h in blocking buffer.

7. Incubate the membrane with the primary antibodies at 4 °C overnight on a shaker.

8. Wash the PVDF membrane for 3 × 10 min in PBST buffer.

9. Incubate the PVDF membrane with fluorescence secondary antibody for 1 h at room temperature.

10. Wash for 3 × 10 min in PBST and detect bands using the Odyssey Imaging software (Image Studio).

4 Notes

1. DEPC-treated RNase-free water should be avoided to dissolve oligonucleotides. Dissolved AONs should be aliquoted and stored at −20 °C and avoid repeated freeze-thaw.

2. For TOES design, the most efficient binding sites of the antisense domain will be the validated exonic or intronic splicing silencers. For exonic silencers, the binding site is favorable to the upstream of the exon.

3. If the antisense domain anneals to an exon, it should avoid inducing any potential exon skipping of the binding exon.

4. Chemical modification of all the RNA nucleotides through the entire tail domain may reduce the binding affinity to protein. However, chemical modification may be only added to the last five nucleotides at the 5′-end of the tail domain (cap, as shown in Fig. 1) to improve the stability while still keep its binding affinity.

5. The duration of transfection can be prolonged to overnight or 24 h. For cells less tolerant to lipofectamine transfection, shorter incubation period, e.g. 6 h, is recommended.

6. It is recommended at least two housekeeping genes are used in the quantitative real-time PCR assay.

Acknowledgments

This work was supported by the Wellcome Trust, University College London, UK Medical Research Council (MRC), SMA-Europe, SMA Trust, Muscular Dystrophy UK and NIHR Great Ormond Street Hospital and Institute of Child Health Biomedical Research Centre.

References

1. Dominski Z, Kole R (1993) Antisense oligonucleotides. Proc Natl Acad Sci U S A 90: 8673–8677

2. Mitrpant C, Porensky P, Zhou H, Price L, Muntoni F, Fletcher S et al (2013) Improved antisense oligonucleotide design to suppress aberrant SMN2 gene transcript processing: towards a treatment for spinal muscular atrophy. PLoS One 8:2–11

3. Hua Y, Sahashi K, Rigo F, Hung G, Horev G, Bennett F et al (2011) Peripheral SMN restoration is essential for long-term rescue of a severe spinal muscular atrophy mouse model. Nature 478:123–126

4. Zhou H, Janghra N, Mitrpant C, Dickinson R, Anthony K, Price L et al (2013) A novel morpholino oligomer targeting ISS-N1 improves rescue of severe spinal muscular atrophy transgenic mice. Hum Gene Ther 24:331–342

5. Singh NK, Singh NN, Androphy EJ, Eliot J, Singh RN (2006) Splicing of a critical exon of human Survival Motor Neuron is regulated by a unique silencer element located in the last intron. Mol Cell Biol 26:1333–1346

6. Mercuri E, Darras BT, Chiriboga CA, Day JW, Campbell C, Connolly AM et al (2018) Nusinersen versus sham control in later-onset spinal muscular atrophy. N Engl J Med 378:625–635

7. Finkel RS, Mercuri E, Darras BT, Connolly NL, Kuntz J, Kirschner CA et al (2017) Nusinersen versus sham control in infantile-onset spinal muscular atrophy. N Engl J Med 377:1723–1732

8. Hoy SM (2017) Nusinersen: first global approval. Drugs 77:473–479

9. Cirak S, Arechavala-Gomeza V, Guglieri M, Feng L, Torelli S, Anthony K et al (2011) Exon skipping and dystrophin restoration in patients with Duchenne muscular dystrophy after systemic phosphorodiamidate morpholino oligomer treatment: an open-label, phase 2, dose-escalation study. Lancet 378:595–605

10. Kinali M, Arechavala-Gomeza V, Feng L, Cirak S, Hunt D, Adkin C et al (2009) Local restoration of dystrophin expression with the morpholino oligomer AVI-4658 in Duchenne muscular dystrophy: a single-blind, placebo-controlled, dose-escalation, proof-of-concept study. Lancet Neurol 8:918–928

11. Aartsma-Rus A, Corey DR (2020) The 10th oligonucleotide therapy approved: golodirsen for Duchenne muscular dystrophy. Nucl Acids Ther 30:67. https://doi.org/10.1089/nat.2020.0845

12. Frank DE, Schnell FJ, Akana C, El-Husayni SH, Desjardins CA, Morgan J et al (2020) Increased dystrophin production with golodirsen in patients with Duchenne muscular dystrophy. Neurology 94:e2270. https://doi.org/10.1212/WNL.0000000000009233

13. Heo YA (2020) Golodirsen: first approval. Drugs 80:329–333

14. Skordis LA, Dunckley MG, Yue B, Eperon IC, Muntoni F (2003) Bifunctional antisense oligonucleotides provide a trans-acting splicing enhancer that stimulates SMN2 gene expression in patient fibroblasts. Proc Natl Acad Sci U S A 100:4114–4119

15. Cartegni L, Krainer AR (2003) Correction of disease-associated exon skipping by synthetic exon-specific activators. Nat Struct Biol 10:120–125

16. Owen N, Zhou H, Malygin AA, Sangha J, Smith LD, Muntoni F et al (2011) Design principles for bifunctional targeted oligonucleotide enhancers of splicing. Nucl Acids Res 39:7194–7208

17. Brosseau JP, Lucier JF, Lamarche AA, Shkreta L, Gendron D, Lapointe E et al (2014) Redirecting splicing with bifunctional oligonucleotides. Nucl Acids Res 42:e40

18. Baughan TD, Dickson A, Osman EY, Lorson CL (2009) Delivery of bifunctional RNAs that target an intronic repressor and increase SMN levels in an animal model of spinal muscular atrophy. Hum Mol Genet 18:1600–1611

19. Osman EY, Yen PF, Lorson CL (2012) Bifunctional RNAs targeting the intronic splicing silencer N1 increase SMN levels and reduce disease severity in an animal model of spinal muscular atrophy. Mol Ther 20:119–126

20. Meyer K, Marquis J, Trüb J, Nlend R, Verp S, Ruepp MD et al (2009) Rescue of a severe mouse model for spinal muscular atrophy by U7 snRNA-mediated splicing modulation. Hum Mol Genet 18:546–555

13

Generation of Human iPSC-Derived Myotubes to Investigate RNA-Based Therapies In Vitro

Pablo Herrero-Hernandez, Atze J. Bergsma and W. W. M. Pim Pijnappel

Abstract

Alternative pre-mRNA splicing can be cell-type specific and results in the generation of different protein isoforms from a single gene. Deregulation of canonical pre-mRNA splicing by disease-associated variants can result in genetic disorders. Antisense oligonucleotides (AONs) offer an attractive solution to modulate endogenous gene expression through alteration of pre-mRNA splicing events. Relevant in vitro models are crucial for appropriate evaluation of splicing modifying drugs. In this chapter, we describe how to investigate the splicing modulating activity of AONs in an in vitro skeletal muscle model, applied to Pompe disease. We also provide a detailed description of methods to visualize and analyze gene expression in differentiated skeletal muscle cells for the analysis of muscle differentiation and splicing outcome. The methodology described here is relevant to develop treatment options using AONs for other genetic muscle diseases as well, including Duchenne muscular dystrophy, myotonic dystrophy, and facioscapulohumeral muscular dystrophy.

Key words Splicing, Human iPSC, Skeletal muscle, Antisense oligonucleotides, In vitro models

1 Introduction

Pre-mRNA splicing is a highly conserved process in eukaryotes that plays a role in pre-mRNA processing. Alternative splicing can diversify gene function to produce isoforms with specific functions in distinct cell types [1, 2]. Genetic variations can lead to defects in pre-mRNA splicing that cause human disease [3]. Modulation of pre-mRNA splicing can be directed to correct aberrant splicing, to skip protein coding variants, to restore the reading frame, or to prevent expression of toxic gene products. This is possible by targeting antisense oligonucleotides (AONs) toward canonical splice sites or to *cis*-acting regulatory elements such as cryptic splice sites or splicing silencers/enhancers [4]. Alternative splicing in skeletal muscle is abundant and essential for muscle development and function [5]. Deregulation of pre-mRNA splicing in skeletal muscle is known to be the underlying cause of multiple human myopathies [5]. Suitable in vitro and in vivo models are crucial to investigate novel splicing modulating drugs in target cells and tissues.

In vitro human skeletal muscle models can be obtained directly from muscle biopsies, or these can be generated by (*trans-*)differentiation of primary fibroblasts, pluripotent stem cells, or non-muscle cells with myogenic capacity like pericytes and mesoangioblasts [6–8]. Several protocols have been described to generate muscle progenitor cells (MPCs) derived from human patient-derived induced pluripotent stem cells (hiPSCs) using directed differentiation methods for disease modeling [9–11].

Here we describe how purified, expandable hiPSC-derived MPCs, generated using a transgene-free procedure [11] can be differentiated into multinucleated myotubes to test the modulating activity of AONs. These methods can be used to analyze splicing correction in vitro to develop RNA-based therapies for muscle disorders. We have used this strategy to test AONs for Pompe disease [12] and describe the methodology here in detail.

2 Materials

All cell culture work needs to be performed under sterile conditions in safety cabinets. All cell lines should be tested for mycoplasma following the manufacturer instructions (Lonza; LT07-318). Cell lines are cultured at 5% CO_2 and 37 °C in humidified incubators.

2.1 Skeletal Muscle Progenitor Cell Culture

1. Human MPC lines (*see* **Note 1**).
2. DMEM 4.5 g/L Glucose.
3. Fetal Bovine Serum.
4. Penicillin/Streptomycin/Glutamine 100× (p/s/g).
5. Fibroblast Growth Factor 2 (FGF2) (*see* **Note 2**).
6. Sterile cell culture grade Bovine Serum Albumin (7.5% BSA).
7. TrypLE™ Express Enzyme (1×), phenol red.
8. Phosphate Buffered Saline (DPBS).
9. Extracellular Matrix gel from Engelbreth (ECM; 1×).
10. DMEM:F12.
11. Insulin/Transferrin/Selenium 100×.
12. DMSO.
13. Freezing containers.

2.2 Cell Culture Media

1. Proliferation medium: DMEM 4.5 g/L Glucose, supplemented with 10% FBS, 1× Pen/Strep, and 100 ng/ml FGF2 (added directly to plate/well).

2. Differentiation medium: DMEM:F12, supplemented with $1\times$ ITS-X and $1\times$ Pen/Strep.

2.3 Antisense Oligonucleotide Design and Delivery

1. Phosphorodiamidate morpholino oligomer (PMO) AONs (Gene Tools, LLC.)

2. Endoporter (Gene Tools, LLC.)

3. MilliQ filtered sterile water.

2.4 Immunofluorescence

1. 4% Paraformaldehyde (diluted from a 32% solution in PBS).

2. 0.1% Tween (diluted in PBS from a 100% solution).

3. 0.3% Triton-X100 (diluted in PBS from a 100% solution).

4. Bovine Serum Albumin (BSA).

5. Primary antibodies: Mouse-α-MYH1E (1:50, MF20 supernatant, DSHB), Rabbit-α-MYOGENIN (1:100, sc-576, Santa Cruz), Rabbit-α-MYOD (1:100, sc-304, Santa Cruz), Mouse-α-PAX7 (1:100, concentrate, DSHB).

6. Secondary antibodies: Horse-α-mouse biotin (1:250, Vector Laboratories), Alexa Fluor-594-a-goat, Alexa Fluor-488-α-mouse, Alexa Fluor-594-α-rabbit, Alexa Fluor-488-α-rabbit (1:500, Invitrogen).

7. Tertiary: Streptavidin 594 (1:500, Invitrogen, S-32356).

8. Hoechst 33342 (1:15,000, Invitrogen, H3570).

9. Nikon wide field microscope ($10\times$ and $20\times$ objectives).

2.5 RNA Isolation, cDNA Synthesis, and Quantitative RT-PCR (RT-qPCR)

1. RNA isolation kit.

2. cDNA Synthesis kit.

3. iTaq Universal SYBR Green Supermix.

4. Hard-Shell 96-Well PCR Plates.

5. Thermocycler.

6. Real-time thermocycler.

7. Spectrophotometer.

8. Agarose.

9. Ethidium Bromide.

10. Primers (*see* Table 1).

Table 1
Primers used for RT-qPCR

Primer target	Sequence (5′ to 3′)
MYOD fw	CACTCCGGTCCCAAATGTAG
MYOD rv	TTCCCTGTAGCACCACACAC
MYOG fw	CACTCCCTCACCTCCATCGT
MYOG rv	CATCTGGGAAGGCCACAGA
LAMP1 fw	GTGTTAGTGGCACCCAGGTC
LAMP1 rv	GGAAGGCCTGTCTTGTTCAC
LAMP2 fw	CCTGGATTGCGAATTTTACC
LAMP2 rv	ATGGAATTCTGATGGCCAAA

3　Methods

3.1　Expansion, Cryopreservation, and Differentiation of MPCs

3.1.1　Expansion

1. MPCs grow optimally when the confluency is between 30 and 90%. It is important to maintain this cell density throughout the expansion (*see* **Note 3**).

2. Plate cells onto ECM coated plates (*see* **Note 4**). Coat plates using a solution of ECM (1:200) diluted in Proliferation medium without FGF2. Coating solution is left on the plates for 30 min at RT.

3. For cell detachment, first wash plates in pre-warmed PBS at 37 °C and then treat the cells with a 1:1 pre-warmed solution of TrypLE™ Express Enzyme and PBS (3 ml for 10 cm plates) for 3–5 min at 37 °C.

4. Collect cells using 5 volumes of Proliferation medium and centrifuge for 4 min at $200 \times g$.

5. Resuspend cells using Proliferation medium, transfer to pre-coated plates (remove coating solution, do not wash), and add 100 ng/ml of FGF2 directly into the plate (*see* **Note 5**).

6. Immediately transfer plates to a humidified incubator and perform cross movements to ensure appropriate cell spreading and mixing of FGF2.

3.1.2　Freeze-Thaw

1. Thaw vials of MPCs in a pre-warmed water bath, transfer cell suspension slowly into 5 volumes of Proliferation medium (no FGF2), and centrifuge for 4 min at $200 \times g$.

2. Plate cells in pre-coated plates using Proliferation medium plus 100 ng/ml of FGF2 freshly added to the cells.

3. Freeze cells using Proliferation medium (plus 100 ng/ml FGF2) and 10% DMSO in 1 ml cryovials and store in freezing containers at −80 °C for 24 h (at least) prior to long-term storage in liquid nitrogen tanks.

3.1.3 Differentiation into Multinucleated Myotubes

1. Grow cells to reach >90% confluency (avoid 100% confluency) and then switch to Differentiation medium for 4 days without refreshing (*see* **Note 6**). Wide field images of differentiated myotubes are shown in Fig. 1.

3.2 Delivery and Efficacy of Antisense Oligonucleotides in Patient-Derived Myotubes

3.2.1 Transfection

1. Resuspend the PMO AONs in RNAse-free MilliQ at a concentration of 1 mM.

2. Add 4.5 μl of Endoporter reagent per ml of medium directly to the cells and mix by gentle shaking (*see* **Note 7**).

3. Add the desired amount of PMO AONs to the cells and mix by gentle shaking.

4. Transfect AONs 1 day prior differentiation (day −1). Cells should be 60–80% confluent.

5. Switch to differentiation medium (day 0).

6. Leave cells to differentiate for 4 days and either collect protein or RNA or fix cells for immunofluorescence.

3.2.2 Immunofluorescence

1. For immunofluorescence analysis of patient-derived myotubes, prepare cells using 48-well plates.

2. Wash cells once in PBS.

3. Fix cells using 4% PFA in PBS for 10 min at RT, remove and add PBS. Cells can be stored at 4 °C before proceeding.

4. Wash twice in PBS for 2 min each.

5. Incubate for 10 min with 0.3% Triton-X100 in PBS for permeabilization.

6. Incubate for 30 min with 3% BSA, 0.1% Tween in PBS for blocking.

7. Repeat washing **step 4**.

8. Incubate with primary antibodies for 1 h at RT in 0.1% BSA, 0.1% Tween in PBS (*see* **Note 8**).

9. Repeat washing **step 4**.

10. Incubate with secondary antibodies for 45 min at RT in 0.1% BSA, 0.1% Tween in PBS.

11. Repeat washing **step 4**.

12. If biotinylated antibodies were used, incubate with tertiary for 30 min at RT in 0.1% BSA, 0.1% Tween in PBS.

13. Repeat washing **step 4**.

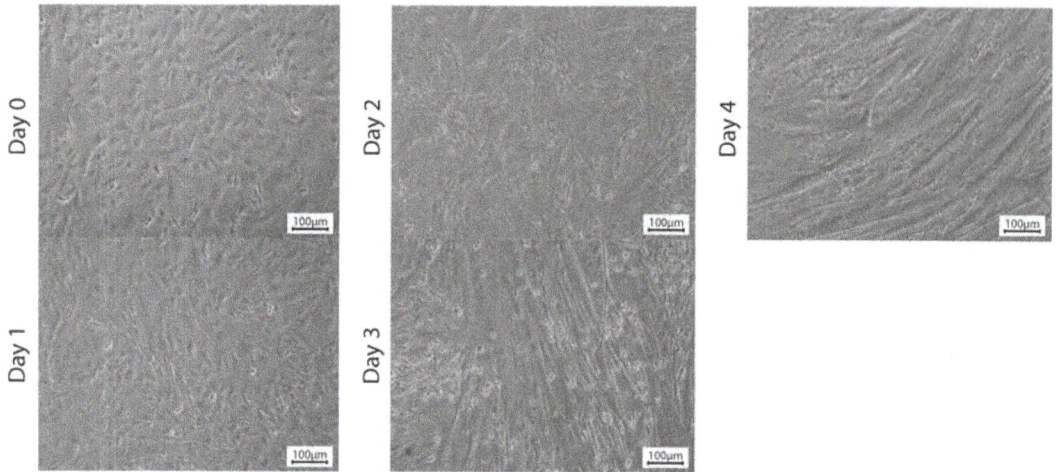

Fig. 1 Wide field images of differentiating MPCs. Representative images of the differentiation of MPCs over 4 days. Scale bar 100 μm

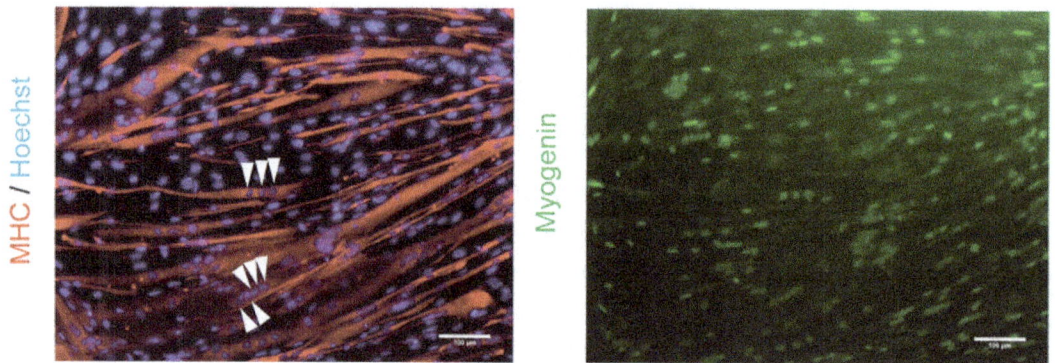

Fig. 2 Immunofluorescence images of 4 days differentiated MPCs. MPCs were stained with MYH1E (red), MYOGENIN (green), and the nuclei with Hoechst (blue). Arrowheads indicate nuclei present in multinucleated myotubes. Scale bar 100 μm

14. Counterstain with Hoechst nuclear staining (1:15,000) in PBS for 10 min.

15. Remove and add PBS.

16. Take images of five random fields with 10× or 20× lens in MYH1E-stained myotubes (*see* Fig. 2) to calculate the fusion index. The fusion index is determined as the percentage of nuclei present in multinucleated MYH1E-positive cells (>2 nuclei in one cell) with respect to the total number of nuclei.

3.2.3 RT-qPCR

1. Harvest RNA after 4 days of differentiation using 350 μl of the lysis buffer (or the amount indicated in the first step of the RNA isolation kit) per well of a 12 well-plate.

2. Purify RNA following the instructions of the preferred RNA isolation kit.

3. Retrotranscribe 300–500 ng of total RNA into cDNA using a cDNA Synthesis Kit.

4. Dilute cDNA samples 10× and prepare the qPCR using iTaq Universal SYBR Green Supermix.

5. Amplify the cDNA of interest using a Real Time System.

6. To analyze alternative splice variants in patient-derived myotubes, we normalize gene expression using each of the following four genes: *MYOD*, *MYOG (Myogenin)*, *LAMP1*, and *LAMP2* (*see* **Note 9**). Gene expression is calculated using the ΔCt method for each housekeeping gene. Thereafter, the average value of the four normalized expression values is calculated.

4 Notes

1. Here we only used transgene-free derived muscle progenitor cells from hiPSCs as described in [11]. However, we anticipate that other sources of myogenic cells would also be applicable.

2. The FGF2 stock powder is dissolved in 0.1% BSA (sterile cell culture-grade BSA diluted in PBS and filtered using a 0.22-μm filter) and aliquoted using tips and tubes that were coated with 0.2% BSA in PBS. The dissolved FGF2 can be stored at −80 °C. Each aliquot is used for maximally 1 week after thawing (kept at 4 °C) and 100 ng/ml is added directly to the cell culture medium every 2 days. When adding FGF2 every 2 days, this can be done without refreshing cell culture media. However, cell culture media must be refreshed every 3 days.

3. MPCs spontaneously differentiate at a confluency of >90% and loose proliferative capacity in culture.

4. Here we only used ECM for our studies. However, we anticipate that other coating materials can be used as well.

5. We typically plate a 1/4 or 1/6 dilution of cells to get a 60–90% confluency in 2 days and 3 days, respectively, using the same plate surface area.

6. There are different methods to differentiate MPCs into multinucleated myotubes [9–12]. For these studies we used DMEM:F12, 1× ITS-X, 1× p/s/g for 4 days without refreshing. Longer differentiation periods might result in cell detachment due to spontaneous contraction.

7. Endoporter reagent is specifically designed for transfection of PMO AONs and was used by us in the following studies [4, 11]. Endoporter reagent does not form a complex with AONs, but it enhances endocytosis in cells. The amount of

Endoporter used is independent of the concentration of PMO AONs. Other backbones might require different delivery reagents.

8. The following proteins are commonly used to assess myogenic potential of differentiating MPCs across species: Myosin heavy chain (MYH1E, cytoplasmic), Myogenin (MYOG, nuclear), and MYOD (nuclear). PAX7 (nuclear) can be used to identify the muscle stem cell fraction.

9. To normalize gene expression, we observed that the following genes involved in myogenesis: *MYOD* and *MYOG*; and the following involved in lysosome biogenesis: *LAMP1* and *LAMP2*; do not change expression levels among patient and healthy donor derived myotubes. We used these genes in this study [12].

Acknowledgments

We thank Erik van der Wal for the critical reading and the revisions of this manuscript. This work was funded by Tex net, the Prinses Beatrix Spierfonds/Stichting Spieren voor Spieren (grant W. OR13-21), the Sophia Children's Hospital Foundation (SSWO) (grant S-687 and S17-32), and Metakids (grant 2016-063).

References

1. Shapiro MB, Senapathy P (1987) RNA splice junctions of different classes of eukaryotes: sequence statistics and functional implications in gene expression. Nucleic Acids Res 15(17): 7155–7174. https://doi.org/10.1093/nar/15.17.7155

2. Lei Q, Li C, Zuo Z, Huang C, Cheng H, Zhou R (2016) Evolutionary insights into RNA trans-splicing in vertebrates. Genome Biol Evol 8(3):562–577. https://doi.org/10.1093/gbe/evw025

3. Bergsma AJ, van der Wal E, Broeders M, van der Ploeg AT, Pim Pijnappel WWM (2018) Alternative splicing in genetic diseases: improved diagnosis and novel treatment options. Int Rev Cell Mol Biol 335:85–141. https://doi.org/10.1016/bs.ircmb.2017.07.008

4. van der Wal E, Bergsma AJ, Pijnenburg JM, van der Ploeg AT, Pijnappel W (2017) Antisense oligonucleotides promote exon inclusion and correct the common c.-32-13T>G GAA splicing variant in Pompe disease. Mol Ther Nucleic Acids 7:90–100. https://doi.org/10.1016/j.omtn.2017.03.001

5. Pistoni M, Ghigna C, Gabellini D (2010) Alternative splicing and muscular dystrophy. RNA Biol 7(4):441–452. https://doi.org/10.4161/rna.7.4.12258

6. Baghdadi MB, Tajbakhsh S (2018) Regulation and phylogeny of skeletal muscle regeneration. Dev Biol 433(2):200–209. https://doi.org/10.1016/j.ydbio.2017.07.026

7. Ito N, Kii I, Shimizu N, Tanaka H, Takeda S (2017) Direct reprogramming of fibroblasts into skeletal muscle progenitor cells by transcription factors enriched in undifferentiated subpopulation of satellite cells. Sci Rep 7(1): 8097. https://doi.org/10.1038/s41598-017-08232-2

8. Selvaraj S, Kyba M, Perlingeiro RCR (2019) Pluripotent stem cell-based therapeutics for muscular dystrophies. Trends Mol Med 25(9): 803–816. https://doi.org/10.1016/j.molmed.2019.07.004

9. Choi IY, Lim H, Estrellas K, Mula J, Cohen TV, Zhang Y, Donnelly CJ, Richard JP, Kim YJ, Kim H, Kazuki Y, Oshimura M, Li HL, Hotta A, Rothstein J, Maragakis N, Wagner KR, Lee G (2016) Concordant but varied

phenotypes among Duchenne muscular dystrophy patient-specific myoblasts derived using a human iPSC-based model. Cell Rep 15(10):2301–2312. https://doi.org/10.1016/j.celrep.2016.05.016

10. Hicks MR, Hiserodt J, Paras K, Fujiwara W, Eskin A, Jan M, Xi H, Young CS, Evseenko D, Nelson SF, Spencer MJ, Handel BV, Pyle AD (2018) ERBB3 and NGFR mark a distinct skeletal muscle progenitor cell in human development and hPSCs. Nat Cell Biol 20(1):46–57. https://doi.org/10.1038/s41556-017-0010-2

11. van der Wal E, Herrero-Hernandez P, Wan R, Broeders M, In 't Groen SLM, van Gestel TJM, van Ijcken WFJ, Cheung TH, van der Ploeg AT, Schaaf GJ, Pijnappel W (2018) Large-scale expansion of human iPSC-derived skeletal muscle cells for disease modeling and cell-based therapeutic strategies. Stem cell Rep 10(6):1975–1990. https://doi.org/10.1016/j.stemcr.2018.04.002

12. van der Wal E, Bergsma AJ, van Gestel TJM, In 't Groen SLM, Zaehres H, Arauzo-Bravo MJ, Scholer HR, van der Ploeg AT, Pijnappel W (2017) GAA deficiency in Pompe disease is alleviated by exon inclusion in iPSC-derived skeletal muscle cells. Mol Ther Nucleic Acids 7:101–115. https://doi.org/10.1016/j.omtn.2017.03.002

14

Determination of Optimum Ratio of Cationic Polymers and Small Interfering RNA with Agarose Gel Retardation Assay

Omer Aydin, Dilek Kanarya, Ummugulsum Yilmaz and Cansu Ümran Tunç

Abstract

Nanomaterials have aroused attention in the recent years for their high potential for gene delivery applications. Most of the nanoformulations used in gene delivery are positively charged to carry negatively charged oligonucleotides. However, excessive positively charged carriers are cytotoxic. Therefore, the complexed oligonucleotide/nanoparticles should be well-examined before the application. In that manner, agarose gel electrophoresis, which is a basic method utilized for separation, identification, and purification of nucleic acid molecules because of its poriferous nature, is one of the strategies to determine the most efficient complexation rate. When the electric field is applied, RNA fragments can migrate through anode due to the negatively charged phosphate backbone. Because RNA has a uniform mass/charge ratio, RNA molecules run in agarose gel proportional according to their size and molecular weight. In this chapter, the determination of complexation efficiency between cationic polymer carriers and small interfering RNA (siRNA) cargos by using agarose gel electrophoresis is described. siRNA/cationic polymer carrier complexes are placed in an electric field and the charged molecules move through the counter-charged electrodes due to the phenomenon of electrostatic attraction. Nucleic acid cargos are loaded to cationic carriers via the electrostatic interaction between positively charged amine groups (N) of the carrier and negatively charged phosphate groups (P) of RNA. The N/P ratio determines the loading efficiency of the cationic polymer carrier. In here, the determination of N/P ratio, where the most efficient complexation occurs, by exposure to the electric field with a gel retardation assay is explained.

Key words Small interfering ribonucleic acid (siRNA), Agarose gel retardation assay, siRNA/cationic polymer carrier complex, Nanoparticles, N/P ratio, Gene delivery

1 Introduction

Regulation of a specific gene has been used for the treatment of a wide range of diseases such as cardiovascular diseases [1], neurodegenerative diseases [2], and cancer [3]. RNA Interference (RNAi) has become a powerful tool for gene silencing studies due to its advantageous properties such as high specificity, effectiveness, a minimum amount of side effect, and easy preparation [4]. RNAi mechanism was first determined by Andrew Z. Fire, Craig

C. Mello, and their colleagues [5]. As a result of their studies, they received the Nobel Prize in Physiology or Medicine in 2006. Interfering RNAs have the ability to silence target genes in cells [6]. At this silencing process, 18–31 nucleotides length small RNA molecules are introduced into cells and induce a sequence-specific gene silencing at the post-transcriptional level by blocking mRNAs containing a matched sequence.

siRNA is the most commonly used interfering RNA and has shown high potential as a therapeutic RNA for gene-based treatments [7]. It regulates the expression of various genes by binding to mRNAs in the cell cytoplasm and causing degradation of their mRNA target. The siRNA is double stranded in nature and is about 22 nucleotides in length. Its precursor is initially recognized by Dicer RNase and then incorporated with the RNA-induced silencing complex (RISC). The siRNA-RISC complex can bind the targeting region of the mRNA and lead to a sequence-specific cleavage with endonuclease Argonaute-2 (AGO2), thereby reducing the expression of the targeted protein [8].

Although siRNA has had particular interest in research, there are some limitations. The major limitations of siRNAs-based therapeutics are their rapid degradation by serum nucleases, poor cellular uptake due to the negatively charged backbone, rapid renal clearance following systemic delivery, off-target effects, and induction of immune responses [9]. In addition, even after siRNA is released from the endosome without being exposed to the lysosome and released into the cell cytoplasm, gene silencing of the siRNA may not be immediately observed [10]. Thereby, before the silencing therapeutic effect of siRNA begins, there is always an induction period due to the intracellular half-life of the target protein. The silencing effect of the given siRNA decreases over time owing to the natural degradation of the siRNA molecules in the cell. Moreover, the therapeutic effect persists for a limited time in rapidly dividing cells such as cancer cells, due to the continuous dilution of siRNA in the replication.

Bare siRNA molecules have poor cellular internalization and they need a carrier to enter cells, where their mechanism of action occurs. The major challenge in the delivery of nucleic acids is the availability of a suitable carrier for transferring siRNA to target cells. To do that, there are two main approaches such as viral and non-viral vectors [11]. The suitable vectors should provide a high degree of transfection for a long period without causing systemic toxicity and immunogenicity [12]. Despite the high transfection activity of viral vectors, possible damage to host genes, immune system stimulation, and infection potential limit its application for gene therapy [13].

To overcome these limitations different types of delivery systems have been designed such as lipid [14, 15], polymer [16], peptide [17, 18], dendrimer [19–21], and micelle [22] based vehicles. An ideal siRNA delivery system should be non-toxic, safe, and effective. Thus, many studies have focused on the development of non-viral vectors with minimal toxicity. Furthermore, the carrier systems should assure entrance of siRNA cargos to the cell cytoplasm without being interrupted by biological barriers such as serum, cell membrane, and endosome/lysosome [23]. In detail, the cell entry of siRNA/cationic polymer carriers is mostly facilitated by the mechanism known as "endocytosis" [10]. In particular, siRNA/cationic polymer carriers should be able to have endosomal escape. Otherwise, siRNA could be degraded in the acidic and enzymatic milieu of endosome/lysosome [24].

Cationic polymers and lipids have frequently been employed in research due to their advantages in gene delivery such as biodegradability, low cytotoxicity, structure variety, and easy scale-up production [25]. Therapeutic nucleic acid cargoes are loaded into the carrier systems mostly through the positive–negative charge interactions between positive charges of carrier and negative charges of phosphate groups in RNA. However, cationic polymer carriers with excess of positive charge may cause toxicity. Cationic carriers cause considerable disruption of cellular membrane integrity because of the negatively charged constitution of cell membrane [26]. Moreover, cationic nanocarriers induce cell necrosis due to the positive charge [27]. They also cause mitochondrial and lysosomal damage and formation of a high number of autophagosomes [28]. In order to overcome this problem, smart carriers have been developed not to cause cell membrane hydrolysis and necrosis so that they can deliver the therapeutic agents to the target site [29]. Although the benefits of nanocarriers in drug delivery have attracted much attention and great efforts have been made to investigate better cationic carriers, toxicity has always been the main problem of cationic carrier applications [30]. Because of this toxicity issue, the number of positive charges of polymer carriers should be kept low. In that case, the required therapeutic concentration of siRNA could not be achieved. Thus, N/P ratio has a great importance for gene delivery. Consequently, it is essential to keep the N/P ratio low, which indicates the complexation efficiency of the anionic therapeutic agents and the cationic carriers, in order to prevent cytotoxicity from excess amount of cationic polymer carriers. As a result, the main aim for optimum N/P ratio is to carry the most efficient number of siRNA with minimum number of cationic polymer carriers.

To identify the optimum ratio of N and P, the gel retardation assay commonly used for nucleic acid separation could be chosen for determination of N/P ratio [31]. This technique is frequently utilized for the determination of DNA or RNA fragments based on their molecular weight [32]. The phosphate backbone of the DNA or RNA is negatively charged, and therefore RNA fragments migrate to the positively charged anode when placed in the electric field for separation, identification, and purification [33].

Agarose which, is a pure linear polymer obtained from seaweed, is frequently used for gel electrophoresis [34]. The polymer is boiled to dissolve in a buffer solution and polymerized in gel form by hydrogen bonding when left to cool at room temperature. There is no other component such as catalysts required than agarose. Therefore, preparing agarose gel is simple and fast. The advantages of agarose gel electrophoresis are being non-toxic gel medium, rapid, and easy to cast of gels and providing well separation of high molecular weight nucleic acids [35]. In addition, the samples can be recovered from the gel by melting or digesting the gel with any agarose enzyme or by treating a chaotropic salt [36].

The movement of molecules in an agarose gel depends on their size, charge, the type of electrophoresis buffer, and the pore size of the gel. In this method, siRNA is forced to migrate through an extremely cross-linked agarose base in response to an electric field. In the solution, the phosphate groups of the siRNA are negatively charged so the siRNA molecule migrates to the positive pole (Fig. 1).

This technique is also being used for determination of the complexation efficiency of siRNA/cationic polymer carriers. By adding a positively charged polymer to the siRNA, the overall charge of the complex is neutralized. Because of decreasing the negative charge density of siRNA complex, the movement of siRNA/cationic polymer carriers in the gel is getting difficult. If the complexation does not happen completely, for example, there are free forms of siRNA with siRNA/cationic polymer carriers, the free siRNAs travel far away than the complexation forms [37].

All in all, the movement of a siRNA/cationic polymer carrier complexation through a gel depends on (a) size of the complexation structure, (b) agarose concentration, (c) type of agarose, (d) applied voltage, (e) presence of staining dye, and (f) electrophoresis buffer type [38]. After running the samples in a suitable dye-containing gel, the complexed and free siRNA can be visualized under UV light.

2 Materials

1. siRNA.
2. Agarose.
3. Tris-HCl.
4. Acetic acid or boric acid.
5. EDTA.
6. DNA Ladder.

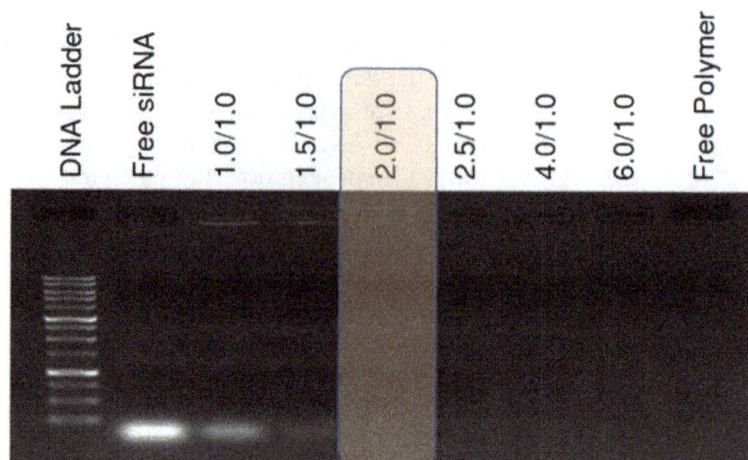

Fig. 1 Agarose gel retardation assay to evaluate siRNA binding efficiency of 15 kDa cationic polymer complexation with siRNA at different N/P (amine/phosphate) ratios. The gel retardation assay is set as the number of carriers is increasing, siRNA is kept constant. The orange-colored rectangle marking in the figure shows the optimal N/P ratio (2/1) which is siRNA completely complexed with the carriers

7. Loading Dye.

8. RedSafe nucleic acid staining solution.

9. Ethidium bromide.

10. dH$_2$O.

11. Microwave oven.

12. Gel casting device.

13. Glass beaker.

14. Graduated cylinder.

Prepare solutions with distilled water and nuclease-free water. Store all reagents (except siRNA and loading dye) at room temperature.

2.1 Agarose Gel

1. Preparation of tris acetate-ethylenediaminetetraacetic (TAE) buffer reagents: Prepare 50× TAE buffer (pH 8.0) with 40 mM ethylenediaminetetraacetic (EDTA) disodium salt (M_W: 336.21 g/mol), 2 mM tris base (M_W: 121.14 g/mol), 1 M Acetic Acid (M_W: 60.05 g/mol) (*see* **Note 1**).

2. Preparation of stock solution 5.0 l 50× TAE buffer (pH 8.0): Weight 1.21 g of Tris base, 67.24 g of EDTA, and draw 285.95 mL of acetic acid and dissolve all of them in 5.0 L distilled water, carefully. Perform this step on the magnetic stirrer and adjust pH 8.0 with hydrochloric acid (HCl) (*see* **Note 2**). Store the solution at room temperature.

3. Preparation of 1× TAE buffer: Draw 20.0 mL of 50× TAE buffer solution and complete it to final 1000.0 mL in a glass beaker (*see* **Note 3**).

4. Preparation of 1% agarose gel: Weight 1.0 g agarose powder, add 100.0 mL of electrophoresis buffer. Agarose powder is dissolved in the electrophoresis buffer to the desired concentration (*see* **Note 4**).

2.2 Polymer and siRNA

1. Preparation of polymer solution: Add 1.0 mg of polymer in 1.0 mL of nuclease-free water. Store at 4 °C.

2. Stock siRNA solution: Dissolve siRNA in nuclease-free water at a concentration of 50 μmol. Afterwards, allocate 800.0 μL of stock siRNA solution into microcentrifuge tubes (*see* **Note 5**) and store at −20 °C.

3 Methods

3.1 Calculation of Nitrogen to Phosphate (N/P) Ratio for Complexation for siRNA/Cationic Polymer Carriers

Following the steps below, the number of siRNA, cationic polymers, and N/P ratios will be calculated. It is good to create a table containing the amounts of siRNA, polymer, loading dye, and distilled water, which will be helpful for setting the experiment (*see* Table 1). After the calculation, the complexation siRNA/cationic polymer carriers at the different N/P ratios are loaded into gel wells, run in the electrophoresis, and visualized under UV light.

To sum up, the procedure can be simply separated into four steps: (1) preparation of materials to be loaded into wells and carried out of complexation, (2) preparation of agarose gel and loading of samples into wells, (3) running the samples at appropriate voltage and time, and (4) obtained data and identification of the optimum N/P ratio (Fig. 2).

3.1.1 Calculation of the Amine Groups of Polymer

1. Calculate the molecular weight and the number of amine groups of the cationic polymer (or look at the datasheet of your polymer if you purchase from a vendor) (*see* **Note 6**), i.e.;

   (a) The molecular weight of a cationic polymer: 180,000.0 g/mol.

   (b) Amine groups per the cationic polymer: 235 amines.

2. Determine the initial dose of the cationic polymer to be used and calculate the mole of the cationic polymer, i.e.;

   (a) The cationic polymer dose: 1.0 mg.

   (b) The number of the cationic polymer moles: 5.5×10^{-9} mol.

Table 1
The siRNA/cationic polymer carrier complexation for different N/P ratios

Well number	Treatment (N/P)	Stock siRNA solution (μL)	Stock polymer solution (μL)	Loading dye (μL)	Nuclease free H₂O (μL)
1	DNA ladder	–	–	–	–
2	Free siRNA	1.0	–	4.0	15.0
3	1.0/1.0	1.0	1.6	4.0	13.4
4	1.5/1.0	1.0	2.4	4.0	12.6
5	2.0/1.0	1.0	3.2	4.0	11.8
6	2.5/1.0	1.0	4.0	4.0	11.0
7	4.0/1.0	1.0	6.4	4.0	8.6
8	6.0/1.0	1.0	9.6	4.0	5.4
9	Free polymer	–	10.6	4.0	5.4

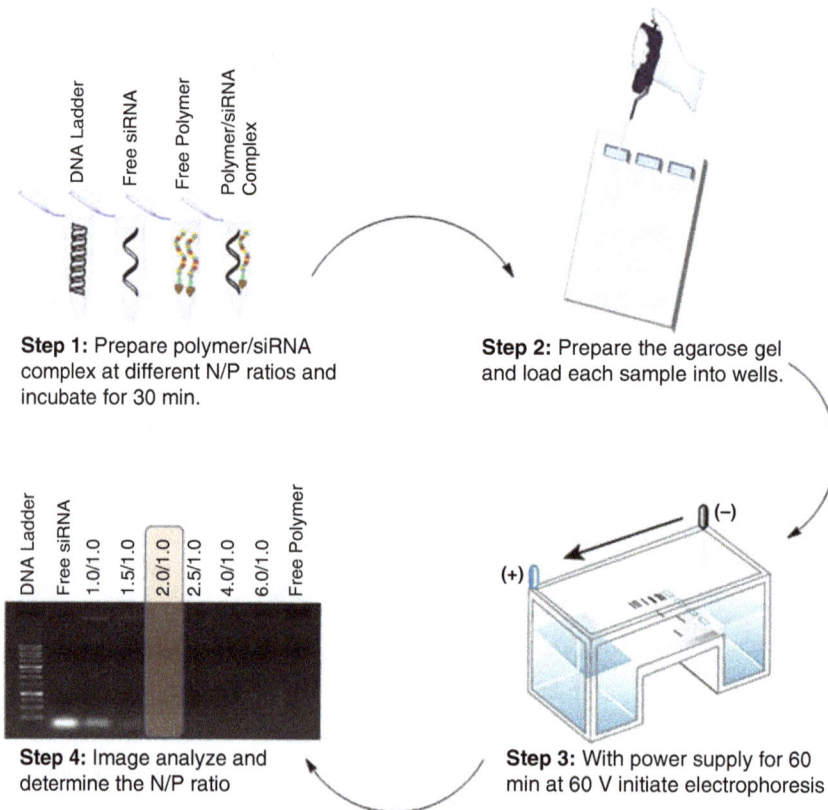

Fig. 2 Brief summary of the creation stages of gel retardation assay

3. Multiply the number of amine groups in the cationic polymer by the number of cationic polymer moles. This number will be the number of total amine moles.

   (a) The mole of amine groups: 1.3×10^{-6} mol.

3.1.2 Calculation of Phosphate Groups of siRNAs

1. Calculate the average molecular weight of siRNAs (*see* **Note 7**), i.e.;

   (a) The average molecular weight of a siRNA molecule: 340.0 g/mol.

   (b) siRNA consists of double stranded 21 bp so it has a total of 42 bp unit/siRNA molecule.

   (c) The molecular weight of 21 bp siRNA (siRNA molecule * number of bp * 2) 14,280.0 g/mol (340.0 * 21 * 2). This is the value used in the subsequent calculations.

2. Determine the initial dose of the siRNA to be used and calculate the mole of the siRNA.

   (a) The siRNA dose: 0.7 μg.

   (b) The number of siRNA moles: 4.9×10^{-11} mol.

3. Multiply the number of Phosphate (PO_4^{3-}) groups in the siRNA to get the total number of phosphate groups.

   (a) The mole of PO_4^{3-} groups (siRNA moles * nr PO_4^{3-} groups): 2.06×10^{-9} mol (4.9×10^{-11} mol * 42).

3.1.3 Calculation of N/P Ratios

One milligram of cationic polymer (1.3×10^{-6} mol obtained from Subheadings 3.1.1–3.1.3) dissolved in 1.0 mL of distilled water as a stock solution. For the calculation of 1:1 N/P ratio, calculate the amount of amine groups in the polymer which equals to the same mole of PO_4^{3-} groups (2.1×10^{-9} mol obtained from the Subheadings 3.1.2 and 3.1.3). Based on this calculation, 1.60 μL of the polymer solution is taken. For 1.5:1, 2:1, 2.5:1, 4:1, 6:1 N/P ratios, this volume is multiplied with 1.5, 2.0, 2.5, 4.0, and 6.0, respectively (*see* Table 1).

3.2 Preparation of 1% Agarose Gel Electrophoresis

1. To prepare 1× TAE buffer, get 50.0 mL of 50× TAE buffer stock solution and complete to 1000.0 mL volume with distilled water in a beaker.

2. Add 1.0 g of agarose powder into 100.0 mL of 1× TAE buffer and mix it to make 1% agarose gel. After that, boil the agarose mixture in a microwave oven to melt it (*see* **Note 8**).

3. Cool the solution to approximately 50–60 ° C.

4. Add the RedSafe nucleic acid staining solution (5.0 μL) (*see* **Note 9**).

5. Poured into a casting tray containing a sample comb and allowed to solidify at room temperature (*see* **Note 10**).

6. Remove the sample comb from the cooling casting tray (*see* **Note 11**).

**3.3 Loading siRNA/
Cationic Polymer
Carrier Complex into
1% Agarose Gel**

1. Prepare mixture of nuclease-free water, polymer, and siRNA in separated microcentrifuge tubes as given in Table 1.

2. Allow 30 min for the complexation of siRNA/cationic polymer carrier.

3. Add 4.0 μL of DNA gel loading dye to each centrifuge tube. The loading dye (6×) is added to the gel for the visualization of siRNA during sample loading and running.

4. Load a total 20.0 μL volume from each sample into agarose gel wells.

5. Set power supply at 60 V for 60 min (*see* **Note 12**).

6. The electrophoretic mobility of the siRNAs of each sample is visualized under UV light (*see* Fig. 1).

4   Notes

1. Agarose gel electrophoresis of siRNA is performed using either TAE buffer or tris-borate–EDTA (TBE) buffer [39].

2. Wear a mask and gloves when preparing buffer agents. TAE buffer solution can be stored at room temperature for a month. In our laboratory, we prepare fresh solution every month.

3. The most reliable strategy for TAE usage, prepare a fresh 1× TAE buffer for each time just before the experiment.

4. The percentage of agarose gels used is generally in the range of 0.2–3%. This percentage could be changed based on the size of siRNA and total molecular weight of the carrier [40].

5. Allocate all the samples (i.e. siRNA) to avoid damaging the main source.

6. Calculation/determination of the molecular weight, chemical formulation, surface charge, and size of the polymer are carried out according to ^{1}H-NMR, zeta potential, and size analysis.

7. The average molecular weight of siRNA base is 340 g/mol. Average molecular weight of a double-stranded DNA (dsDNA) base pair and a single-stranded DNA (ssDNA) base pair are 600 and 330 g/mol, respectively [41].

8. Initially, the agarose solution is boiled for 45–60 s in the microwave at 180 °C. If the agarose powder does not dissolve well, repeat the step for another 20–25 s until it totally dissolves.

9. Instead of using RedSafe nucleic acid staining dye, ethidium bromide, bromophenol blue or xylene cyanole could be preferred [35].

10. The casting tray containing sample comb must be on a flat surface. Otherwise, the thickness of the gel may not be uniform. While pouring the solution, one should be careful about the formation of bubbles, which may affect the migration efficiency.

11. To solidify the gel, wait 1–1.5 h before removing the comb from the tray to ensure that the loading wells are intact and well-opened.

12. The optimum running time is 45–90 min to identify the N/P ratio of siRNA/cationic polymer carrier complexation. In the case of gel electrophoresis is performed at a higher voltage, the running time is decreased. However, if the running time is increased that leads to heating up which leading to siRNA banding artifacts [39].

Acknowledgments

This work was supported by Scientific and Technological Research Council of Turkey (TUBITAK)-2515 COST Program Grant number: 118Z952 and Research Fund of Erciyes University (Project number: MAP-2020-9692).

References

1. Zhang X, Li DY, Reilly MP (2019) Long intergenic noncoding RNAs in cardiovascular diseases: challenges and strategies for physiological studies and translation. Atherosclerosis 281:180–188. https://doi.org/10.1016/j.atherosclerosis.2018.09.040

2. Conejos-Sanchez I, Gallon E, Nino-Pariente A, Smith JA, De la Fuente AG, Di Canio L, Pluchino S, Franklin RJM, Vicent MJ (2019) Polyornithine-based polyplexes to boost effective gene silencing in CNS disorders. Nanoscale 12(11):6285–6299. https://doi.org/10.1039/c9nr06187h

3. Liu T, Wang L, Xin H, Jin L, Zhang D (2020) Delivery systems for RNA interference-based therapy and their applications against cancer. Sci Adv Mater 12(1):75–86. https://doi.org/10.1166/sam.2020.3694

4. Subhan MA, Torchilin VP (2019) Efficient nanocarriers of siRNA therapeutics for cancer treatment. Transl Res 214:62–91. https://doi.org/10.1016/j.trsl.2019.07.006

5. Fire A, Xu S, Montgomery MK, Kostas SA, Driver SE, Mello CC (1998) Potent and specific genetic interference by double-stranded RNA in Caenorhabditis elegans. Nature 391(6669):806–811. https://doi.org/10.1038/35888

6. Dong Y, Siegwart DJ, Anderson DG (2019) Strategies, design, and chemistry in siRNA delivery systems. Adv Drug Deliv Rev 144:133–147. https://doi.org/10.1016/j.addr.2019.05.004

7. Kapadia CH, Luo B, Dang MN, Irvin-Choy ND, Valcourt DM, Day ES (2020) Polymer nanocarriers for MicroRNA delivery. J Appl Polym Sci 137(25):48651. https://doi.org/10.1002/app.48651

8. Lin Y-X, Wang Y, Blake S, Yu M, Mei L, Wang H, Shi J (2020) RNA nanotechnology-mediated cancer immunotherapy. Theranostics 10(1):281–299. https://doi.org/10.7150/thno.35568

9. Karimi F, Azizi Jalilian F, Hossienkhani H, Ezati R, Amini R (2019) siRNA delivery technology for cancer therapy: promise and challenges. Acta Med Iran 57:83–93. https://doi.org/10.18502/acta.v57i2.1760

10. Tsouris V, Joo MK, Kim SH, Kwon IC, Won Y-Y (2014) Nano carriers that enable co-delivery of chemotherapy and RNAi agents for treatment of drug-resistant cancers. Biotechnol Adv 32(5):1037–1050. https://doi.org/10.1016/j.biotechadv.2014.05.006

11. Shirley JL, de Jong YP, Terhorst C, Herzog RW (2020) Immune responses to viral gene therapy vectors. Mol Ther 28(3):709–722. https://doi.org/10.1016/j.ymthe.2020.01.001

12. Sabouri-Rad S, Oskuee RK, Mahmoodi A, Gholami L, Malaekeh-Nikouei B (2017) The effect of cell penetrating peptides on transfection activity and cytotoxicity of polyallylamine. Bioimpacts 7(3):139–145. https://doi.org/10.15171/bi.2017.17

13. Rezaee M, Oskuee RK, Nassirli H, Malaekeh-Nikouei B (2016) Progress in the development of lipopolyplexes as efficient non-viral gene delivery systems. J Control Release 236:1–14. https://doi.org/10.1016/j.jconrel.2016.06.023

14. Parashar D, Rajendran V, Shukla R, Sistla R (2020) Lipid-based nanocarriers for delivery of small interfering RNA for therapeutic use. Eur J Pharm Sci 142:105159. https://doi.org/10.1016/j.ejps.2019.105159

15. Jorge A, Pais A, Vitorino C (2020) Targeted siRNA delivery using lipid nanoparticles. Methods Mol Biol 2059:259–283

16. Liao X (2019) Development of a cationic polymer-based siRNA delivery system for the regeneration of tendon injury. University of East Anglia, Norwich

17. Qiu Y, Tam B, Chung WY, Mason J, Lam JK (2019) Relationship between the secondary structure of the peptide base siRNA carrier and effective gene silencing effect on lung epithelial cells. J Aerosol Med Pulmonary Drug Delivery 2:A16–A17

18. Cummings JC, Zhang H, Jakymiw A (2019) Peptide carriers to the rescue: overcoming the barriers to siRNA delivery for cancer treatment. Transl Res 214:92–104. https://doi.org/10.1016/j.trsl.2019.07.010

19. Pan J, Mendes LP, Yao M, Filipczak N, Garai S, Thakur GA, Sarisozen C, Torchilin VP (2019) Polyamidoamine dendrimers-based nanomedicine for combination therapy with siRNA and chemotherapeutics to overcome multidrug resistance. Eur J Pharm Biopharm 136:18–28. https://doi.org/10.1016/j.ejpb.2019.01.006

20. Laurini E, Marson D, Aulic S, Fermeglia M, Pricl S (2019) Evolution from covalent to self-assembled PAMAM-based dendrimers as nanovectors for siRNA delivery in cancer by coupled in silico-experimental studies. Part II: self-assembled siRNA Nanocarriers. Pharmaceutics 11(7):324. https://doi.org/10.3390/pharmaceutics11070324

21. Jain A, Mahira S, Majoral J-P, Bryszewska M, Khan W, Ionov M (2019) Dendrimer mediated targeting of siRNA against polo-like kinase for the treatment of triple negative breast cancer. J Biomed Mater Res A 107(9):1933–1944. https://doi.org/10.1002/jbm.a.36701

22. Lu Y, Zhong L, Jiang Z, Pan H, Zhang Y, Zhu G, Bai L, Tong R, Shi J, Duan X (2019) Cationic micelle-based siRNA delivery for efficient colon cancer gene therapy. Nanoscale Res Lett 14(1):193. https://doi.org/10.1186/s11671-019-2985-z

23. Kandil R, Xie Y, Heermann R, Isert L, Jung K, Mehta A, Merkel OM (2019) Coming in and finding out: blending receptor-targeted delivery and efficient endosomal escape in a novel bio-responsive siRNA delivery system for gene knockdown in pulmonary T cells. Adv Ther 2(7):1900047. https://doi.org/10.1002/adtp.201900047

24. Prantner AM, Scholler N (2014) Biological barriers and current strategies for modifying nanoparticle bioavailability. J Nanosci Nanotechnol 14(1):115–125. https://doi.org/10.1166/jnn.2014.8899

25. Patil S, Gao Y-G, Lin X, Li Y, Dang K, Tian Y, Zhang W-J, Jiang S-F, Qadir A, Qian A-R (2019) The development of functional non-viral vectors for gene delivery. Int J Mol Sci 20(21):5491. https://doi.org/10.3390/ijms20215491

26. Chen J, Hessler JA, Putchakayala K, Panama BK, Khan DP, Hong S, Mullen DG, Dimaggio SC, Som A, Tew GN, Lopatin AN, Baker JR, Holl MMB, Orr BG (2009) Cationic nanoparticles induce nanoscale disruption in living cell plasma membranes. J Phys Chem B 113(32):11179–11185. https://doi.org/10.1021/jp9033936

27. Wei X, Shao B, He Z, Ye T, Luo M, Sang Y, Liang X, Wang W, Luo S, Yang S, Zhang S, Gong C, Gou M, Deng H, Zhao Y, Yang H, Deng S, Zhao C, Yang L, Qian Z, Li J, Sun X, Han J, Jiang C, Wu M, Zhang Z (2015) Cationic nanocarriers induce cell necrosis through impairment of Na+/K+-ATPase and cause subsequent inflammatory response. Cell Res

25(2):237–253. https://doi.org/10.1038/cr.
2015.9

28. Fröhlich E (2012) The role of surface charge in cellular uptake and cytotoxicity of medical nanoparticles. Int J Nanomedicine 7: 5577–5591. https://doi.org/10.2147/IJN.S36111

29. Durmaz YY, Lin Y-L, ElSayed MEH (2013) Development of degradable, pH-sensitive star vectors for enhancing the cytoplasmic delivery of nucleic acids. Adv Funct Mater 23(31): 3885–3895. https://doi.org/10.1002/adfm.201203762

30. Lv H, Zhang S, Wang B, Cui S, Yan J (2006) Toxicity of cationic lipids and cationic polymers in gene delivery. J Control Release 114(1): 100–109. https://doi.org/10.1016/j.jconrel.2006.04.014

31. Scott V, Clark AR, Docherty K (1994) The gel retardation assay. In: Protocols for gene analysis. Springer, Berlin, pp 339–347

32. Hellman LM, Fried MG (2007) Electrophoretic mobility shift assay (EMSA) for detecting protein–nucleic acid interactions. Nat Protoc 2(8):1849

33. Helling RB, Goodman HM, Boyer HW (1974) Analysis of endonuclease R-EcoRI fragments of DNA from lambdoid bacteriophages and other viruses by agarose-gel electrophoresis. J Virol 14(5):1235–1244

34. Magdeldin S (2012) Gel electrophoresis: principles and basics. BoD–Books on Demand, Norderstedt

35. Chawla H (2002) Introduction to plant biotechnology. Routledge, Milton Park. https://doi.org/10.1201/9781315275369

36. Tietz D (2012) Nucleic acid electrophoresis. Springer, Berlin

37. Loening UE (1968) Molecular weights of ribosomal RNA in relation to evolution. J Mol Biol 38(3):355–365. https://doi.org/10.1016/0022-2836(68)90391-4

38. Rio DC, Ares M Jr, Hannon GJ, Nilsen TW (2010) Nondenaturing agarose gel electrophoresis of RNA. Cold Spring Harb Protoc 2010(6):pdb.prot5445. https://doi.org/10.1101/pdb.prot5445

39. Sanderson BA, Araki N, Lilley JL, Guerrero G, Lewis LK (2014) Modification of gel architecture and TBE/TAE buffer composition to minimize heating during agarose gel electrophoresis. Anal Biochem 454:44–52. https://doi.org/10.1016/j.ab.2014.03.003

40. Murphy D, Carter DA, Evans MJ (1993) Transgenesis techniques: principles and protocols. Springer, Berlin

41. He F (2011) DNA molecular weight calculation. Bio-Protocol 1(6):e46. https://doi.org/10.21769/BioProtoc.46

15

Design and Delivery of SINEUP: A New Modular Tool to Increase Protein Translation

Michele Arnoldi, Giulia Zarantonello, Stefano Espinoza, Stefano Gustincich, Francesca Di Leva and Marta Biagioli

Abstract

SINEUP is a new class of long non-coding RNAs (lncRNAs) which contain an inverted Short Interspersed Nuclear Element (SINE) B2 element (invSINEB2) necessary to specifically upregulate target gene translation. Originally identified in the mouse *AS-Uchl1* (antisense Ubiquitin carboxyl-terminal esterase L1) locus, natural SINEUP molecules are oriented head to head to their sense protein coding, target gene (*Uchl1*, in this example). Peculiarly, SINEUP is able to augment, in a specific and controlled way, the expression of the target protein, with no alteration of target mRNA levels. SINEUP is characterized by a modular structure with the Binding Domain (BD) providing specificity to the target transcript and an effector domain (ED)—containing the invSINEB2 element—able to promote the loading to the heavy polysomes of the target mRNA. Since the understanding of its modular structure in the endogenous *AS-Uchl1* ncRNA, synthetic SINEUP molecules have been developed by creating a specific BD for the gene of interest and placing it upstream the invSINEB2 ED. Synthetic SINEUP is thus a novel molecular tool that potentially may be used for any industrial or biomedical application to enhance protein production, also as possible therapeutic strategy in haploinsufficiency-driven disorders.

Here, we describe a detailed protocol to (1) design a specific BD directed to a gene of interest and (2) assemble and clone it with the ED to obtain a functional SINEUP molecule. Then, we provide guidelines to efficiently deliver SINEUP into mammalian cells and evaluate its ability to effectively upregulate target protein translation.

Key words SINEUP, Long non-coding RNA, Antisense, Translational increase, Physiological increase, Therapeutic tool, Haploinsufficiency, Protein manufacturing

1 Introduction

The quantitative improvement of protein production in mammalian systems is a compelling need for the industrial manufacturing of commercially available enzymes, antibodies and supplements, but also for gene therapy-based treatments of medical conditions. Several technologies are available to address such a need, however, they usually consist of introducing exogenously constructs containing the protein of interest or directly the target peptide [1]. These approaches still struggle to overcome hazardous but invariable hurdles, especially when used as therapeutic tools, such as ectopic expression and protein quantity modulation, sometimes associated with toxicity [2]. As an alternative, newly identified RNA-based techniques such as small activating RNAs (RNAa) are able to target and upregulate endogenous gene transcription [3]. In 2012, Carrieri et al. discovered a new class of lncRNAs, belonging to the category of natural antisense transcripts (NATs), that have the property to increase the protein translation of the target mRNA [4]. These transcripts were named SINEUP based on their ability to upregulate target protein translation by means of an invSINEB2 repeat, leaving unaltered the transcriptional levels of the target mRNA. They were first discovered in mice, where *AS-Uchl1* was found to have a post-transcriptional upregulating activity on its sense protein coding counterpart, *Uchl1* mRNA [4]. Later studies confirmed and validated the expression of SINEUP in human cells [5, 6]. SINEUP molecular mechanism relies on its modular structure, composed of two fundamental domains: a Binding Domain (BD)—a region at the 5' of the lncRNA overlapping head to head to the 5' of the target mRNA—and an Effector Domain (ED)—constituted by an invSINEB2 repetitive element. The BD is crucial for target gene pairing and it confers molecular specificity, while the ED is the functional part of SINEUP required for loading the target transcript on polysomes and driving the translational increase [4].

The initial discovery was then supported by the crucial finding that the BD could be engineered in order to target a specific mRNA of interest, as first demonstrated with Green Fluorescent Protein (*GFP*) [4]. Additionally, miniSINEUP containing only the BD and a shorter version of the original ED were also proven to be effective [7]. This characteristic would enable to overcome the difficulties of long molecules delivery, especially for therapeutic purposes in which naked RNA molecules administration can be proposed. Recently, TranSINE Therapeutics Limited (Cambridge, UK) has been founded to translate the SINEUP technology into the clinics as therapeutics for haploinsufficiency.

All together, these findings qualified SINEUP as a flexible tool able to upregulate the protein production of virtually any mRNA target of interest, affecting solely the translational levels. As such, SINEUP molecules demonstrated to be a suitable tool for protein manufacturing, able to boost, for instance, the production of recombinant proteins and monoclonal antibodies in mammalian cells [8–10]. SINEUP molecules are also being studied from a therapeutic point of view, since their functional characteristics could confer advantages with respect to other gene therapy approaches. SINEUP molecules generally upregulate the endogenous protein of about two- to fivefold (almost within a physiologic range) and they are only effective in those districts where the target mRNA is physiologically expressed, therefore avoiding unspecific effects [10]. Finally, by targeting mRNA molecules, SINEUP does not introduce stable or unwanted changes in the host genome. Thus far, several synthetic SINEUP molecules were successfully designed and delivered as potential therapeutic molecules. For instance, synthetic SINEUP were designed and successfully used to increase the levels of disease-associated proteins in vitro, such as Parkinson's disease-associated DJ-1 in three human neuronal cell lines [7] and Glial-cell derived neurotrophic factor (GDNF) in mouse cell line [11]. Moreover, SINEUP have been used to rescue frataxin levels in a cellular model of Friedreich's Ataxia [12]. Concerning in vivo model systems, SINEUP could effectively rescue some phenotypes associated with microphthalmia with linear skin defects (MLS) syndrome in a medaka fish model of *cox7B* haploinsufficiency [13]. More recently, SINEUP targeting GDNF mRNA was tested in a neurochemical Parkinson's disease (PD) mouse model [11]. Interestingly, SINEUP-GDNF increased endogenous GDNF level for at least 6 months which lead to an enhancement of dopamine release in the striatum and an amelioration of motor behavior and neurodegeneration, without affecting body weight or food intake, common side effects of the ectopic expression of GDNF [11].

Here, we describe in detail the fundamental steps in order to design, clone, and deliver SINEUP molecular tools in the cellular system of interest.

2   Materials

2.1   Design and Cloning of SINEUP

1. Zenbu browser (https://fantom.gsc.riken.jp/zenbu/).

2. Ensembl genome browser (https://www.ensembl.org/index.html).

3. UCSC genome browser (https://genome.ucsc.edu/).

4. NCBI (https://www.ncbi.nlm.nih.gov/genome/).

5. RNA Fold web server (http://rna.tbi.univie.ac.at/cgi-bin/RNAWebSuite/RNAfold.cgi).

6. Salt free primers. Use primers to clone the Effector Domain from Carrieri et al. [4]:

    (a) For mAS *Uchl1*Δ5′: 5- CAGTGCTAGAGGAGGTCA GAAGAG-3

    (b) Rev mAS *Uchl1* fl: 5-CATAGGAGTGTTTCATT-3

    Or primers from Zucchelli et al. [7]:

    (a) FWD EcoRI *invSINEB2*: 5-TATAGAATTCCAGTGCTA GAGGAGG-3

(b) 3REV HindIII *invSINEB2*: 5- GAGAAAGCTTAAGA GACTGGAGC-3

7. Ultrapure water.

8. Mammalian expression plasmid vector of your choice.

9. Restriction enzymes.

10. T4 DNA ligase.

11. *E. coli* competent cells.

12. Luria-Bertani (LB-10 g/L Tryptone, 10 g/L NaCl, 5 g/L Yeast Extract) broth.

13. Antibiotics of your choice.

14. Maxiprep kit.

15. Molecular grade agarose.

16. DNA gel staining.

17. DNA loading dye.

18. DNA ladder.

19. Electrophoresis apparatus.

2.2 SINEUP Delivery into Cellular Model

1. Target cells of interest. In this protocol we used human iPS-derived neuronal progenitor cells (hiNPCs) [14].

2. Poly-L-ornithine hydrobromide (20 μg/mL).

3. Laminin (3 μg/mL).

4. Dulbecco's Modified Eagle's Medium (DMEM).

5. Ham's F-12 Nutrient (HAM F12).

6. B27.

7. Penicillin-Streptomycin solution.

8. L-Glutamine.

9. EGF (20 ng/mL).

10. bFGF (20 ng/mL).

11. Heparin (5 μg/mL).

12. hiNPCs culturing medium: 70% v/v DMEM completed with 30% v/v HAM F12, 2% v/v B27, 1% v/v Penicillin-Streptomycin solution, and 1% v/v L-Glutamine and supplemented with EGF (20 ng/mL), bFGF (20 ng/mL) and Heparin (5 μg/mL). Semi-confluent monolayers of hiNPCs were maintained in 5% CO_2, 37 °C humidified incubator.

13. Enzyme for adherent cellular culture detachment (e.g. TripLE in this example).

14. Transfection solution (e.g. Nucleofector Solutions—Lonza).

15. NucleofectorTM Device (Lonza).

2.3 RNA and Protein
Analysis

1. RadioImmunoPrecipitation Assay—RIPA—buffer.

2. Protease and Phosphatase inhibitors (PI and PhI, respectively).

3. Bicinchoninic acid (BCA) or Bradford protein assay kit.

4. Bovine serum albumin (BSA) standard curve.

5. Western blot apparatus.

6. Specific antibody for the target protein and a housekeeping protein.

7. Enhanced ChemiLuminescence (ECL) or equivalent assay.

8. Eurofins PCR primer design tool or equivalent (https://eurofinsgenomics.eu/en/ecom/tools/pcr-primer-design/).

9. ThermoFisher primer analyzer tool or equivalent (https://www.thermofisher.com/it/en/home/brands/thermo-scientific/molecular-biology/molecular-biology-learning-center/molecular-biology-resource-library/thermo-scientific-web-tools/multiple-primer-analyzer.html).

10. DNAse.

11. DEPC-treated (nuclease-free) water.

12. Retro-transcriptase containing both oligo(dT) and Random Hexamer primers.

13. SYBR Green or Taqman reagents.

3   Methods

Prepare all the solutions using analytical grade reagents with ultra-pure water at room temperature, unless otherwise indicated. Solutions used with cells are filtered or sterilized at the beginning. Cells are handled under biological hoods while all the other reactions are performed on the bench. Follow safety instructions indicated by safety team in your institution.

3.1 Binding Domain
Design and Cloning

1. Retrieve the sequence of your transcript of interest from Zenbu browser (https://fantom.gsc.riken.jp/zenbu/) [15].

2. In Zenbu, select the organism in which you will perform the experiments (e.g. human or mouse) and search for the target gene of interest (*see* **Note 1** and Fig. 1a).

3. Check how many TSS have been characterized for your gene and in which tissue/cell line they are expressed (Fig. 1b) by selecting the "CAGE libraries" (or other libraries in which you are interested).

4. Expand the window in the 5′UTR region containing the TSS in order to better appreciate how many TSS have been identified. You will see a bar plot in which each bar represents a TSS

A

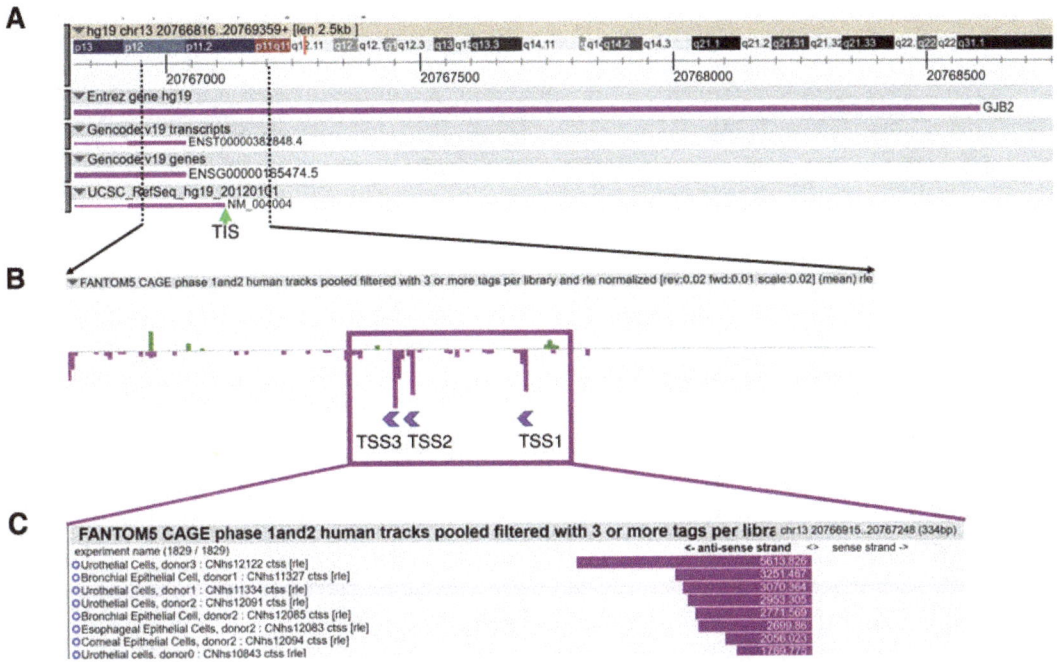

B

C

Fig. 1 Zenbu genome browser interrogation of transcript of interest showing TSS usage in selected model system. (**a**) Zenbu genome browser screenshot showing the genomic location of the gene of interest (in this example *GJB2*) in the human genome assembly 19 (hg19), green arrow identifies the Translation Initiation Site (TIS). RefSeq accession number for the transcript of interest (NM_004004, in this case) is also visible. In the enlargement in (**b**) alternative TSS usage identified by FANTOM5 CAGE library (reported as purple bar plots) is shown. Both TSSs in the same orientation (reverse strand, purple bar plots) and in the opposite orientation (forward strand, green bar plots) of the gene are depicted. The purple box highlights the chosen region in which interrogates the browser about the expression of your target transcript. Main TSSs, TSS1, TSS2, and TSS3, are indicated by purple arrows. (**c**) Displays the tracks (experiments) in which the transcript is more expressed

(Fig. 1b); in the window below, the tracks expressing your transcript will be displayed (Fig. 1c). Among them, you can select the tissue/cells/organ of your interest.

5. In the identified TSS and the 5′UTR region, design an efficient SINEUP molecule in which the BD overlaps the Translational Initiation Site (TIS) and expands in the upstream region characterized for your transcript.

6. Further test the expression of your transcript(s) using genome browsers such as Ensembl genome browser (https://www.ensembl.org/index.html), UCSC genome browser (https://genome.ucsc.edu/), or NCBI (https://www.ncbi.nlm.nih.gov/genome/).

7. Design forward primer (FW) on the 5′UTR and reverse primer (RV) on coding DNA sequence (CDS) downstream Translational Initiation Site (TIS).

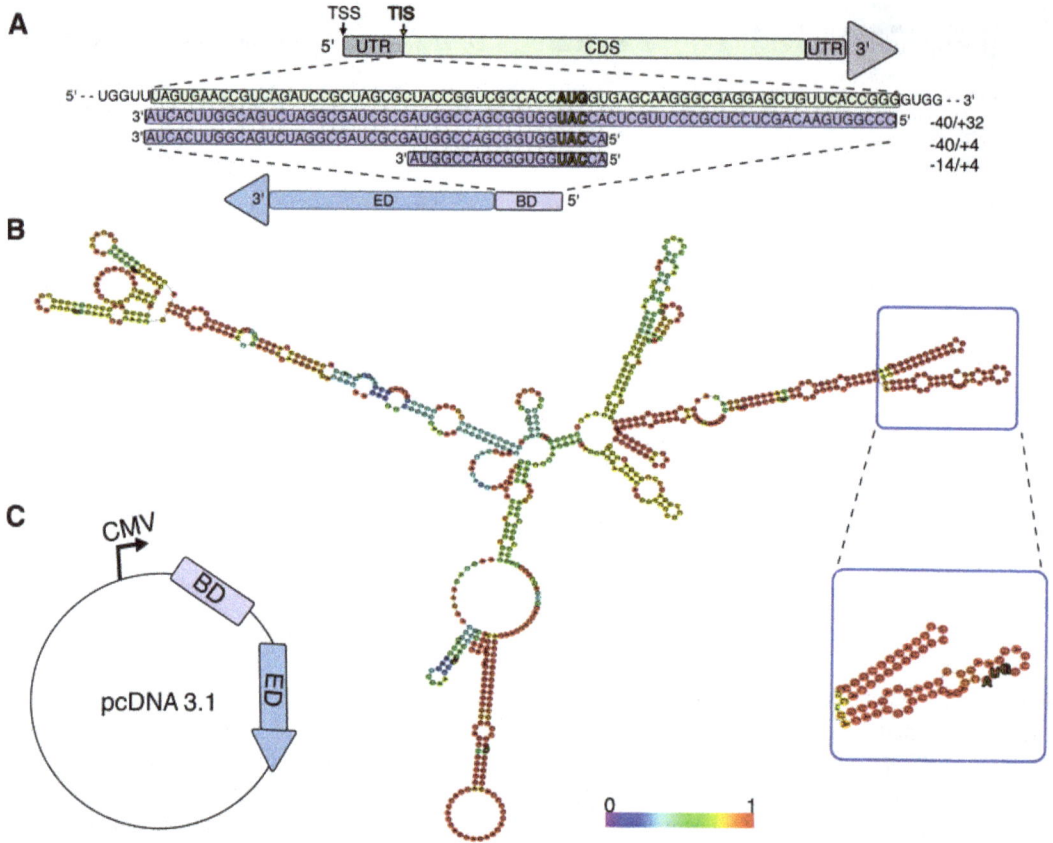

Fig. 2 SINEUP modular structure: Binding domain (BD) design and cloning. (**a**) Representation of the modular structure of SINEUP and the design strategy to obtain BD oriented head to head to its target gene. BDs, in light purple, targeting the gene of interest (as an example, the CDS of EGFP is depicted in light green), untranslated region (UTR) in gray, Transcriptional Start Site (TSS) in black, and Transcriptional Initiation Site (TIS) in yellow, are shown. Different suggested lengths for BD are reported. SINEUP effector domain (ED) in light blue, with the antisense region, overlapping TIS is depicted in black-shadowed yellow. Adenine (A) of the TIS is set as 0, BDs lengths vary between −40 from the A and +32 nucleotides, −40 and +4 nucleotides, and −14 and +4 nucleotides, respectively. (**b**) Example of RNA folding prediction results using RNA Fold web server (http:/rna. tbi.univie.ac.at/cgi-bin/RNAWebSuite/RNAfold.cgi) with default parameters, here reported for EGFP transcript. The blue box displays an enlargement of the TIS (in yellow AUG) surrounding region, in which packed structures without big hairpin-loop are reported. Minimum free energy (MFE) prediction is used to create the output, with base-pair probability in form of a dot plot it depicted. Scale bar reports both base-pair and unpair probability for every base colored dot, with 1 highest probability that the base pair as well as 1 highest probability that the base unpair with the neighbors (**c**). (Schematic representation of pcDNA 3.1 vector structure obtained from Carrieri C. et al. 2012 [4], BD is depicted in light purple followed by ED in light blue are under CMV promoter)

8. Dissolve salt free quality primers in ultrapure water to obtain equal molarity for each primer to a final concentration of 10 μM.

9. Confirm the expression of your target transcript(s) by PCR in your selected model system.

10. Run the PCR product onto agarose gel to confirm amplicon size, purify PCR product, and sequence with both FW and RV primers to validate transcript identity and specificity (*see* **Note 2**).

11. The BD will be designed spanning the TIS of the transcript of interest. You can start selecting the "canonical" target sequence corresponding to 40 bases upstream and 4 bases downstream the TIS ($-40/+4$) in which 0 correspond to the A of the TIS ATG (*see* **Note 3**).

12. Additional sequences of different length can be designed both on the TIS and in region(s) targeting downstream, in-frame methionines (*see* **Note 4** and Fig. 2a). A full list of published BDs is reported in Table 1.

13. On the target region of interest where the BD will be designed, evaluate the GC content and the mRNA secondary structure using RNA Fold web server (http://rna.tbi.univie.ac.at/cgi-bin/RNAWebSuite/RNAfold.cgi—Fig. 2b, *see* **Note 5**) [19, 20].

14. Further analysis can be performed to understand if other lncRNA(s) are generated endogenously from the same genetic locus. Albeit not mandatory, this information can give you an idea of the specific gene's structure as well as highlight possible competition in SINEUP binding to the target TIS.

15. Once selected BDs of interest, perform off-targets prediction by interrogating Basic local Alignment (BLAST, https://blast.ncbi.nlm.nih.gov/Blast.cgi) or BLAT (https://genome.ucsc.edu/cgi-bin/hgBlat). This information can help to avoid SINEUP off-target issues and to prioritize experiments with the more specific molecules (*see* **Note 6**).

16. When you select the region of the transcript in which you want to design the BD, you must reverse complement the sense mRNA sequence to obtain a SINEUP able to interact with the target gene (Fig. 2a).

17. For each BD of interest, synthetize the FW primer in a sense orientation ($5'$–$3'$ orientation) with the restriction enzyme needed for insertion into the chosen vector at its $5'$ end. On the other hand, synthetize the RV primer in an inverted orientation ($3'$-$5'$) with the compatible restriction site at the $5'$ end, in order to allow primer annealing.

18. Equimolar concentration of both primers must be denatured for $5'$ at 100 °C and cooled slowly (1 °C every minute) to obtain double stranded DNA (dsDNA) BD.

Table 1
List of natural and synthetic SINEUP tested to upregulate protein production. Binding Domain (BD—A and B columns, the overlapping sequence with the sense spliced mRNA is reported) and Effector Domain (ED—C column) administered with different delivery methods (F columns) in diverse model systems (H and I columns) are listed. Position relatively to the targeted Methionine (A column) of the gene of interest (E column) are reported (AUG with A set as 0, M1 refers to first Methionine corresponding to TIS, whereas M76 refers to Methionine in position 76 in the amino acid chain)

BD length	BD specificity	Class of ED	Target gene	Delivery method	Model organism	Increase level of target protein	Reference
−40/−49 M1, −9/+34 M1	Uchl1	invSINEB2/Alu	Uchl1	Lipofectamine 2000	MN9D	Yes	[4]
−40/+32 M1	Uchl1	invSINEB2/Alu	Uchl1	Lipofectamine 2000	HEK293T/17	Yes	[4]
−54/−6 M1, +65/+98 M1	Uxt	invSINEB2/Alu	Uxt	Lipofectamine 2000	MN9D	Yes	[4]
−40/+32 M1	GFP	invSINEB2/Alu	GFP	Lipofectamine 2000	HEK293T/17	Yes	[4]
−40/+32 M1	FLAG	invSINEB2	FLAG-TRAF6, FLAG-DJ-1, FLAG-Hba, FLAG-TTRAP	Lipofectamine 2000	HEK293T/17	Yes	[7]
−40/+32 M1	DJ1	invSINEB2	DJ1	Lipofectamine 2000	SH-SY5Y, BE(2)-M17 and SK-N-SH	Yes	[7]
−40/+4 M1	DJ1	invSINEB2	DJ1	Lipofectamine 2000	SH-SY5Y	Yes	[7]

(continued)

Table 1
(continued)

BD length	BD specificity	Class of ED	Target gene	Delivery method	Model organism	Increase level of target protein	Reference
−40/+32 M1	EGFP	invSINEB2 or invSINEB2/Alu	EGFP	Lipofectamine 2000	HEK 293T/17, HepG2, HeLa	Yes	[7]
−40/+32 M1	EGFP	invSINEB2/Alu	EGFP	Lipofectamine 3000	HeLa, HEK293A, HEK293T and CHO-K1	Yes	[16]
−40/+32 M1	EGFP	invSINEB2/Alu	EGFP-BBCK	Lipofectamine 3000	HEK293T	Yes	[16]
−40/+32 M1	EGFP	invSINEB2/Alu	EGFP-βB2-crystallin	Lipofectamine 3000	HEK293T	Yes	[16]
−40/+32 M1	Sox9	invSINEB2/Alu	Sox9	Lipofectamine 3000	C5.18 and rat primary chondrocyte	Yes	[16]
−40/+32 M1	met-luciferase	invSINEB2/Alu	met-luciferase	Lipofectamine 3000	HEK293T	Yes	[16]
−40/0 M1	EGFP	invSINEB2/Alu	EGFP	Lipofectamine 3000	HEK293T	No	[16]
−100/0 M1	EGFP	invSINEB2/Alu	EGFP	Lipofectamine 3000	HEK293T	No	[16]
0/+32 M1	EGFP	invSINEB2/Alu	EGFP	Lipofectamine 3000	HEK293T	No	[16]
0/+100 M1	EGFP	invSINEB2/Alu	EGFP	Lipofectamine 3000	HEK293T	No	[16]
0/+200 M1	EGFP	invSINEB2/Alu	EGFP	Lipofectamine 3000	HEK293T	Yes	[16]

−40/+360 M1	EGFP	invSINEB2/Alu	EGFP	Lipofectamine 3000	HEK293T	Yes	[16]
−40/+560 M1	EGFP	invSINEB2/Alu	EGFP	Lipofectamine 3000	HEK293T	No	[16]
−40/+760 M1	EGFP	invSINEB2/Alu	EGFP	Lipofectamine 3000	HEK293T	No	[16]
−40/+100 M1	EGFP	invSINEB2/Alu	EGFP	Lipofectamine 3000	HEK293T	Yes	[16]
−40/+200 M1	EGFP	invSINEB2/Alu	EGFP	Lipofectamine 3000	HEK293T	Yes	[16]
−100/+32 M1	EGFP	invSINEB2/Alu	EGFP	Lipofectamine 3000	HEK293T	Yes	[16]
−100/+100 M1	EGFP	invSINEB2/Alu	EGFP	Lipofectamine 3000	HEK293T	Yes	[16]
−40/+32 M1	anti-HIV antibody 10E8	invSINEB2 or invSINEB2/Alu	anti-HIV antibody 10E8	Lipofectamine 3000	HEK293T	Yes	[16]
−40/+100 M1	anti-HIV antibody 10E8	invSINEB2 or invSINEB2/Alu	anti-HIV antibody 10E8	Lipofectamine 3000	HEK293T	Yes	[16]
−100/+32 M1	anti-HIV antibody 10E8	invSINEB2 or invSINEB2/Alu	anti-HIV antibody 10E8	Lipofectamine 3000	HEK293T	Yes	[16]
−200/+48 M1	anti-HIV antibody 10E8	invSINEB2 or invSINEB2/Alu	anti-HIV antibody 10E8	Lipofectamine 3000	HEK293T	Yes	[16]
−40/+32 M1	EGFP	invSINEB2/Alu	EGFP	FreeStyle MAX reagent	CHO-S	Yes	[8]

(continued)

Table 1
(continued)

BD length	BD specificity	Class of ED	Target gene	Delivery method	Model organism	Increase level of target protein	Reference
−40/+4 M1	*NLuc*	invSINEB2/Alu	*NLuc*	Lipofectamine 2000; FreeStyle MAX reagent	HEK293T; CHO-S	Yes	[8]
−40/+33 M1	*mIgG* secretory leader	invSINEB2/Alu	*scFv, Periostin*	FreeStyle MAX reagent	CHO-S	Yes	[8]
−13/+82 M1	*mEln* secretory leader	invSINEB2/Alu	*NLuc*	FreeStyle MAX reagent	CHO-S	Yes	[8]
−40/+32 M1	*EGFP*	invSINEB2/Alu	*EGFP*	Lipofectamine 2000	HEK293T/17	Yes	[13]
−40/+32 M1	*EGFP*	invSINEB2/Alu	*EGFP*	RNA injection	Zebrafish embryo	Yes	[13]
−40/+32 M1	*cox7B*	invSINEB2/Alu	*cox7B*	RNA injection	Zebrafish embryo	Yes	[13]
−67/+31 M1	*PPP1R12A*	FRAM	*PPP1R12A*	Lipofectamine 2000	HEK293T, Hela	Yes	[5]
−35/+4 M1	*PPP1R12A*	FRAM	*PPP1R12A*	Lipofectamine 2000	HEK293T	Yes	[5]
−35/+4 M1	*PPP1R12A*	invSINEB2	*PPP1R12A*	Lipofectamine 2000	HEK293T	Yes	[5]
−93/+95 M1	*ITFG2*	MIRb	*ITFG2*	Lipofectamine 2000	HEK293T	Yes	[5]

−28/+32 M1	GFP	invSINEB2/Alu	GFP	Lipofectamine 2000	HEK293T/17	Yes	[17]
−28/+27 M1	GFP	invSINEB2/Alu	GFP	Lipofectamine 2000	HEK293T/17	No	[17]
−28/+22 M1	GFP	invSINEB2/Alu	GFP	Lipofectamine 2000	HEK293T/17	No	[17]
−28/+18 M1	GFP	invSINEB2/Alu	GFP	Lipofectamine 2000	HEK293T/17	No	[17]
−28/+12 M1	GFP	invSINEB2/Alu	GFP	Lipofectamine 2000	HEK293T/17	No	[17]
−28/+8 M1	GFP	invSINEB2/Alu	GFP	Lipofectamine 2000	HEK293T/17	No	[17]
−28/+4 M1	GFP	invSINEB2/Alu	GFP	Lipofectamine 2000	HEK293T/17	Yes	[17]
−28/0 M1	GFP	invSINEB2/Alu	GFP	Lipofectamine 2000	HEK293T/17	No	[17]
−28/−4 M1	GFP	invSINEB2/Alu	GFP	Lipofectamine 2000	HEK293T/17	No	[17]
−28/−8 M1	GFP	invSINEB2/Alu	GFP	Lipofectamine 2000	HEK293T/17	No	[17]
−28/−12 M1	GFP	invSINEB2/Alu	GFP	Lipofectamine 2000	HEK293T/17	No	[17]
−20/+32 M1	GFP	invSINEB2/Alu	GFP	Lipofectamine 2000	HEK293T/17	No	[17]
−10/+32 M1	GFP	invSINEB2/Alu	GFP	Lipofectamine 2000	HEK293T/17	No	[17]
0/+32 M1	GFP	invSINEB2/Alu	GFP	Lipofectamine 2000	HEK293T/17	No	[17]
10/+32 M1	GFP	invSINEB2/Alu	GFP	Lipofectamine 2000	HEK293T/17	No	[17]

(continued)

Table 1
(continued)

BD length	BD specificity	Class of ED	Target gene	Delivery method	Model organism	Increase level of target protein	Reference
−18/+4 M1	GFP	invSINEB2/Alu	GFP	Lipofectamine 2000	HEK293T/17	No	[17]
−40/+4 M1	FXN	invSINEB2/Alu	FXN	Lipofectamine 2000	HEK 293T/17	Yes	[12]
−40/0 M1	FXN	invSINEB2/Alu or invSINEB2	FXN	Lipofectamine 2000	HEK 293T/17	Yes	[12]
−14/0 M1	FXN	invSINEB2/Alu or invSINEB2	FXN	Lipofectamine 2000	HEK 293T/17	Yes	[12]
−14/+4 M1	FXN	invSINEB2/Alu or invSINEB2	FXN	Lipofectamine 2000	HEK 293T/17	Yes	[12]
−40/+4 M76	FXN	invSINEB2/Alu	FXN	Lipofectamine 2000	HEK 293T/17	Yes	[12]
−40/0 M76	FXN	invSINEB2/Alu or invSINEB2	FXN	Lipofectamine 2000	HEK 293T/17	Yes	[12]
−10/−60/+0 M76	FXN	invSINEB2	FXN	Lipofectamine 2000	HEK 293T/17	No	[12]
−40/+4 M1	FXN	invSINEB2	FXN	lentiviral particles	GM04078 fibroblasts	Yes	[12]
−40/0 M1	FXN	invSINEB2	FXN	lentiviral particles	GM04078 fibroblasts	Yes	[12]

−14/0 M1	FXN	invSINEB2	FXN	lentiviral particles	GM04078 fibroblasts; SH-SY5Y	Yes	[12]
−14/+4 M1	FXN	invSINEB2	FXN	lentiviral particles	GM04078 fibroblasts; SH-SY5Y	Yes	[12]
−40/+4 M76	FXN	invSINEB2	FXN	lentiviral particles	GM04078 fibroblasts	Yes	[12]
−40/+4 M1	FXN	invSINEB2	FXN	electroporation	GM16214 lymphoblasts	Yes	[12]
−40/0 M1	FXN	invSINEB2	FXN	electroporation	GM16214 lymphoblasts	Yes	[12]
−40/+4 M1	GDNF	invSINEB2/Alu	GDNF	Lipofectamine 2000; AAV9 injection	Neuro2a; C8-D1A	Yes	[11]
−14/+4 M1	GDNF	invSINEB2/Alu	GDNF	Lipofectamine 2000; AAV9 injection	Neuro2a; C8-D1A	Yes	[11]
−14/+4 M1	GDNF	invSINEB2/Alu	GDNF	AAV9 vectors injection	Dorsal striatum of adult C57BL/6J mice	Yes	[11]
−31/+4 M1	SOX9	invSINEB2/Alu	SOX9	Lipofectamine 2000	HepG2 and Hepa 1-6	Yes	[18]
−28/+4 M1	EGFP	invSINEB2/Alu	EGFP	Lipofectamine 2000	HEK293T/17	Yes	[18]

19. By taking advantage of the restriction sites present in the BD and in the selected plasmid of interest, clone every BD upstream of the *AS-Uchl1* ED. A shorter version of ED called miniSINEUP, which still comprehends the invSINEB2 element present in *AS-Uchl1*, should preferentially be used to obtain shorter SINEUP (*see* **Note 7**) in expression plasmid (e.g. pcDNA 3.1 vector from Clontech, Fig. 2c) using T4 DNA ligase (Fig. 2b). If necessary, inducible or tissue specific promoters can be used.

3.2 Effector Domain Design and Cloning

You can design the ED and insert in a different plasmid as needed. The ED comprehends an invSINEB2, free right Alu monomer (FRAM) or MIRb (Mammalian-wide Interspersed Repeat type b) transposable element sequence from natural SINEUP (*see* **Note 8**). Virtually any of those elements can present SINEUP activity, however, the element present in *AS-Uchl1* was previously inserted in miniSINEUP vectors and it was the most widely characterized. For this reason, in this section, we will refer to invSINEB2 cloning as ED.

1. To amplify ED from *AS-Uchl1* you can use primers "For mAS *Uchl1Δ5'*"and "Rev mAS *Uchl1* fl" from Carrieri et al. [4]. Moreover, to obtain a 170 bp long ED, primers "FWD EcoRI *invSINEB2*" and "REV HindIII *invSINEB2*" from Zucchelli et al. [7] which produce a shorter SINEUP called miniSINEUP, can be selected. Albeit miniSINEUP retains only the invSINEB2 and not the Alu element, this construct was successfully used to increase GFP translation [7].

2. Add restriction enzymes sites at the 5' end to the primers of interest to clone the ED into the desired vector.

3. After performing PCR, purify the amplicon and confirm the correct sequence.

4. Clone the purified ED in the vector of interest downstream the BD.

5. Transform the ligation product into *E. coli* competent cells.

6. Cultured the bacteria in LB broth at 37 °C supplemented with the antibiotic required for selection (e.g. kanamycin sulfate).

7. The empty vector and/or a vector containing only the ED (in the case you clone only the BD targeting the gene of interest in the vector already containing the ED) must be produced as control for ligation.

8. Verify positive clones by restriction analysis or colony PCR and sequencing.

9. Produce and purify the plasmids containing SINEUP for the gene of interest and the empty vector.

10. Analyze plasmids integrity on agarose gel and confirm again the presence of the insert and its correct orientation by restriction map and sequencing.

3.3 SINEUP Delivery into Cellular Model

Different methods such as transfection, electroporation, or viral vectors transduction, can be chosen to deliver your plasmid into the cells of choice. The selected method should meet a criterion of high plasmid internalization efficiency (>60–70% positive cells). In fact, SINEUP molecules upregulate protein translation in a physiologic range (approximately two- to fivefold) so if few cells receive the plasmid this effect can be hidden or under-estimated. Here, we describe the method which best fits our expectations in hiNPCs [14]. For additional information on SINEUP delivery in animal and cellular models *see* **Note 9**.

In this protocol we focused on our target cells of choice, hiNPCs, for which electroporation represents an optimal method of transfection for transient expression. For this purpose, we used Nucleofector™ Device (Lonza) and selected the recommended protocol for our specific cell line, as well as the advised program (A033) of the device. Dealing specifically with SINEUP, we recommend the following:

1. Start by using 1 µg of SINEUP for every one million cells (*see* **Note 10**).

2. Prepare the electroporation mix using the recommended electroporation solution and each SINEUP you want to deliver as well as a solution containing the empty vector as negative control.

3. Resuspend the cellular pellet with the electroporation mix and electroporate with the device and program of choice (e.g. Nucleofector™ Device, Lonza program A033 for hiNPCs).

4. Collect the electroporated cells and seed them in a mix of 1: 2 = old:new pre-warmed medium. We recommend seeding the cells to reach roughly 40% confluency.

5. Replace the medium after 6 h of incubation (be careful that cells are attached on the well) with new, pre-warmed medium.

6. If any fluorescent protein (e.g. GFP) reporter gene is present, check the fluorescence levels to evaluate delivery efficiency (Fig. 3a).

7. We recommend harvesting the electroporated cells after 24–48 h, when cells have fully recovered and are transiently expressing the delivered constructs (containing SINEUP or empty control).

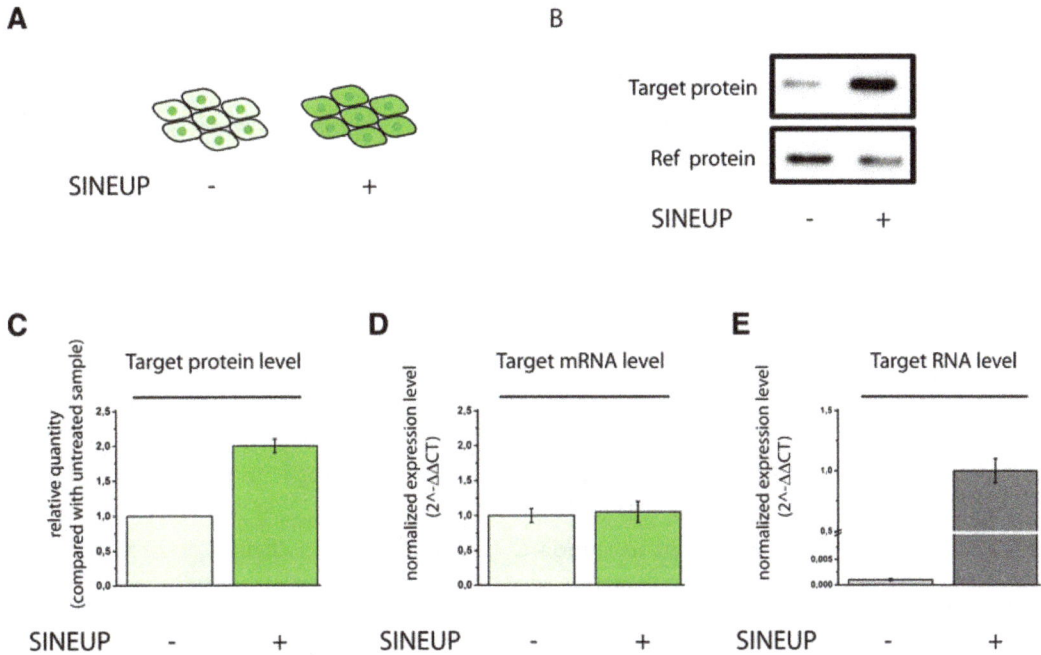

Fig. 3 Expected results when SINEUP positively increases target mRNA translation without altering its transcriptional level in human induced neural progenitor cells (hiNPCs). (**a**) Representative image of hiNPCs expressing EGFP after electroporation with control vector (SINEUP −) or with SINEUP targeting *EGFP* (SINEUP +). (**b**) Western Blot image reports target protein expression relative to a chosen reference protein in the presence (+) or absence (−) of SINEUP against the target mRNA, *EGFP*, in this example. (**c**) Bar graph representing fold change quantification of target protein from WB experiments in (**b**). (**d**) Bar graphs representing normalized expression level of target mRNA upon SINEUP administration (− control SINEUP, + SINEUP against target *EGFP* mRNA). SINEUP presence is depicted in (**e**). Transcripts expression was obtained from qPCR quantification and normalized with a chosen reference gene

3.4 SINEUP Efficacy Assessment: Transcription and Translation Evaluation

Successful SINEUP function needs to be assessed by transcriptional and translational analysis, such as qPCR and WB, respectively. In particular, an increase in protein production between two- and fivefold changes is expected (Fig. 3b, c) without dysregulation of the target mRNA expression (Fig. 3d) when SINEUP is present (Fig. 3e).

1. Apply protocols you routinely use in your lab to extract proteins from the electroporated cells. We used commercial RIPA buffer supplemented with Protease and Phosphatase inhibitors (PI and PhI, respectively).

2. Carefully quantify your proteins using BCA or Bradford protein assay kit. Always include the BSA standard curve for protein quantification. It is fundamental reliable quantification in order to appreciate standard two- to fivefold changes in the protein level due to SINEUP delivery.

3. Perform Western blot experiments using a protocol you are familiar with; the optimization of WB depends on the target protein of interest.

4. Load the same amount of protein (e.g. 20 μg) for all the samples.

5. It is crucial to use specific antibodies to recognize the protein isoform of interest targeted by SINEUP. In particular, the resolved band of the target gene (Fig. 3b) should be clearly distinguishable after WB analysis to perform a correct quantification of the protein of interest (Fig. 3c).

6. Be careful with your choice of ECL or equivalent assay. If the reagent is not sensitive enough you will not see the band of interest; on the other side, many sensitive reagents will not allow you to appreciate subtle changes in protein level due to saturated bands.

7. Albeit in standard experiments three/five biological replicates are considered enough to achieve robust results and to assess reproducibility, in this case you may need more replicates (*see* **Note 11**).

8. Design Quantitative Polymerase Chain Reaction (qPCR) primers for the target of interest, a housekeeping gene and SINEUP ED by using Eurofins PCR Primer Design online tool (https://eurofinsgenomics.eu/en/ecom/tools/pcr-primer-design/), but equivalent systems could also be employed (*see* **Note 12**).

9. All primers should be tested for hairpin and primer dimer formation. Here we used multiple primer analyzer (Thermo-Fisher; https://www.thermofisher.com/it/en/home/brands/thermo-scientific/molecular-biology/molecular-biology-learning-center/molecular-biology-resource-library/thermo-scientific-web-tools/multiple-primer-analyzer.html) and Oligo Calc (Northwestern; http://biotools.nubic.northwestern.edu/OligoCalc.html), but similar tools might also be evaluated.

10. Apply protocols you routinely use in your lab to obtain high quality RNA from the electroporated cells.

11. Preferentially, perform DNase treatment on your RNA samples before proceeding with retro-transcription in order to avoid issues due to DNA contamination.

12. Always resuspend RNA in DEPC-treated (nuclease-free) water.

13. To retro-transcribe RNA to produce cDNA we used a mix containing both oligo(dT) and Random Hexamer primers.

14. For qPCR, the Master Mix reagents should be preferred to DNA polymerase in order to minimize errors. You can use SYBR Green or Taqman systems equally.

15. Perform qPCR assay with the cDNA using primers or Taqman probes targeting a reference housekeeping gene (such as *ACTB*), the target transcript(s) and SINEUP to assess correct expression.

16. Calculate the relative mRNA expression level of the target gene by normalizing treated samples with SINEUP against the empty vector and by using $2^{-\Delta\Delta Ct}$ method [21].

17. Normalize the empty vector against the SINEUP sample for SINEUP expression (*see* **Note 13**).

18. Stable mRNA expression from your target transcript should be observed (Fig. 3d) when SINEUP is expressed (Fig. 3e).

4   Notes

1. The first step is to identify the specific transcript isoform(s) of your transcript of interest that is expressed in the selected model system. This step is necessary since your target might express alternative isoforms that differ from cell to cell and in different tissues. Moreover, the expressed isoform in your cell model/tissue would not necessarily be the most characterized. Zenbu browser is an on-demand freely available interface that allows visualizing data, such as RNA-seq, CAGE (cap analysis of gene expression), short-RNA-seq, and ChIP-seq (chromatin immunoprecipitation) in the chosen model system. The broad spectrum of systems that are annotated could give information especially on annotated promoters and transcriptional starting sites (TSS) usage. However, not all the systems have been characterized and reported in Zenbu [15]. Additional information about how to use Zenbu genome browser, as well as methods to detect and analyze SINEUP in cell culture can be found in Takahashi et al. [22].

2. When you design SINEUP BD, it is important to seek for possible Single Nucleotide Variants (SNV) that might hamper the annealing capability of SINEUP or the expression of your target.

3. Most of the efficient SINEUP BDs characterized until now have been designed in this region in order to specifically recognize the target gene. Specific recognition of transcript originating from the gene of interest allows increase in translation of the full-length target. We recommend designing BD of different lengths. Optimization of the Binding Domain sequence and length is an important step in designing the experiment

[17]. Among others, we suggest −40/+32, −40/+4, and 14/+4 BD regions as the most promising target to be chosen for the experiments (as reported in Fig. 2a).

4. Additional SINEUP targeting in-frame, internal methionine can be design if the TIS surrounding region is poorly accessible. Although many more experimental evidences support the importance of targeting the TIS, the internal, in-frame methionine of the *Frataxin* gene was successfully tested and provided efficient upregulation of Frataxin protein [12]. In fact, SINEUP mechanism is not completely characterized and we cannot exclude that target mRNA can be loaded on polysomes by internal bait which does not interfere with translational initiation. Moreover, alternative translation might occur to downstream AUG codon in the presence of internal ribosomal entry site (IRES) [23]. In summary, we prompt researchers to explore also the possibility of targeting internal in-frame methionines.

5. As mentioned in the methods, packed secondary structures can impair the sense-antisense pairing of SINEUP molecules, decreasing its efficacy. In addition, GC rich region and stem-and-loop structures might inhibit translation by blocking ribosome binding [23]. RNA fold web server predicts in vitro secondary structure of single stranded nucleic acid by energy minimization and/or minimum entropy. This analysis will provide hints on the accessibility of the transcript area in order to optimize SINEUP binding. Although RNA fold tool can be a useful tool, it only mimics the complex physiological condition of RNA fold present in vivo. Recently, a novel approach called icShape (in vivo click selective 2-hydroxyl acylation and profiling experiment) was established to capture the RNA structure in vivo in order to overcome the RNA Fold tool issues [24]. Although not essential for SINEUP design, we recommend analysing icShape data to better understand the binding capacity of SINEUP to the target area of interest.

6. While performing off-target analysis, keep in mind that the possible off-target transcripts need also to be expressed in the model system of your choice to be targeted by SINEUP. On the other hand, the TIS or the internal in-frame methionine you choose to design SINEUP must be present in the transcript(s) expressed in the tissue of interest for translation to occur.

7. To clone SINEUP you can use a mammalian expression plasmid vector of your choice. Since the aim is to obtain an abundant and efficient production of SINEUP, the expression should be driven by a suitable promoter, Cytomegalovirus (CMV) or SV40, frequently employed in mammalian cells plasmids. An inducible/tissue specific promoter may be used, if needed.

8. Long sequences of in vitro synthesized RNA molecules can generate issues especially during delivery in cells. For this reason, starting from the original complete sequence containing the invSINEB2, Alu sequence, and 3′ tail of the natural *AS-Uchl1*, a short functional version was created, called miniSINEUP, encompassing only the invSINEB2 [7, 17]. We suggest performing the SINEUP experiment using miniSINEUP construct, composed by the specific BD and 170 nucleotides length ED of invSINEB2 element. Notably, the human transcriptome does not present invSINEB2 sequences, but functional SINEUP were discovered in human containing embedded FRAM and MIRb transposable element. These two functional domains worked as Eds [5]. Although ED can be designed either with invSINEB2, FRAM, and MIRb domains, we recommend to generate initially ED by using invSINEB2 motif that has been extensively and successfully used to increase protein translation in different cell lines, in vivo in Medaka fish and in a mouse model for Parkinson Disease [4–6, 10–13]. Notably, by nuclear magnetic resonance (NMR) analysis, it was observed that the secondary structure of the invSINEB2 motif is crucial for its function. For this reason, when designing the ED containing invSINEB2, it is important to include the region between 43 and 58 stem-loop structure since this region is likely to be vital for SINEUP function [25].

9. In vitro and in vivo delivery can be optimized and changed if needed. To date, successful SINEUP deliveries were previously reported in vitro using Lipofectamine and lentiviral particles in human derived fibroblasts [12], in vivo by RNA injection in zebrafish embryo [13] and AAV9 vectors injection in the dorsal striatum of adult mice [11]. For both in vitro and in vivo, the delivery method can vary according to the model of choice, the stage and the time window of interest (e.g. transient expression, conditional expression or stable expression). The methods described here for BD design, cloning and SINEUP efficacy assessment steps in cellular models also apply for animal systems. However, concerning the delivery step in vivo, SINEUP molecules can be delivered and expressed in animal models similarly to other RNA-based therapeutics, e.g. by viral vectors. Moreover, it is possible to administer in vitro transcribed SINEUP targeting the transcript of interest instead of the plasmid carrying it. If this method is used, you should chemically modify the in vitro transcribed SINEUP to better stabilize the molecule, by replacing CTP with 5-methylcytidine-50-triphosphate (m5C); and replacing UTP with pseudouridine-50-triphosphate (Ψ) or N1-methylpseudouridine-50-triphosphate (N1mΨ) [18]. In our experience, it is best to test whether the experimental cell system chosen for the

experiment is positively responding to SINEUP administration. For this reason, we suggest testing the previously characterized SINEUP against *GFP* [4, 7], before moving to SINEUP targeting the gene of interest. To analyze the SINEUP efficacy on GFP translation, WB analysis as well as imaging analysis with semi-automatic detection methods can be performed [22]. Of importance, to underlie that protein translation is increased due to SINEUP mechanism, qPCR should be conducted and report stable expression of the target transcript.

10. SINEUP molecule can act in a dose dependent manner to control protein translation without affecting transcript level [7]. To assess the optimal efficiency, different concentrations of SINEUP should be tested.

11. SINEUP delivery experiments as well as WB and qPCR analysis should be performed at least in triplicate to observe statistically significant changes. However, depending on the model system used, additional experiments may be needed to reach significance. The crucial step is the WB quantification: if the band of your target(s) and the reference proteins are not well resolved, changes ascribed to SINEUP positive upregulation can be difficult to examine. In our experience it is better to set specific conditions for WB detection in the chosen model system before performing SINEUP experiment.

12. Primers or Taqman probes for target gene(s) of interest and stable housekeeping gene(s) must be designed in order to assess the stability of the target gene(s) upon SINEUP administration. In addition, primers or probes targeting SINEUP ED must be designed in order to check the correct expression of SINEUP into the chosen model system after plasmid delivery (if the chosen ED contain the invSINEB2 and Alu elements the desired primer can be found in Zucchelli et al. [7]). Primers for qPCR should be designed in regions spanning exon-exon junctions to minimize DNA residual contamination, maximize PCR efficiency and specificity.

13. The different methods for analysing SINEUP and the target mRNA expression is needed since for SINEUP, the quantification in qPCR of the empty vector should be close to the detection limit of the instrument since little signal can be recorded by using the SINEUP specific primer or probes. On the contrary, SINEUP expression is expected to be high upon delivery of the plasmid containing SINEUP sequence, hence that amount is set as default condition one, whereas the empty vector quantification should be close to zero (Fig. 3e).

References

1. Wahlestedt C (2013) Targeting long non-coding RNA to therapeutically upregulate gene expression. Nat Rev Drug Discov 12(6):433–446. https://doi.org/10.1038/nrd4018

2. Wang D, Gao G (2014) State-of-the-art human gene therapy: part I. Gene delivery technologies. Discov Med 18(97):67–77

3. Li LC (2017) Small RNA-guided transcriptional gene activation (RNAa) in mammalian cells. Adv Exp Med Biol 983:1–20. https://doi.org/10.1007/978-981-10-4310-9_1

4. Carrieri C, Cimatti L, Biagioli M, Beugnet A, Zucchelli S, Fedele S, Pesce E, Ferrer I, Collavin L, Santoro C, Forrest AR, Carninci P, Biffo S, Stupka E, Gustincich S (2012) Long non-coding antisense RNA controls Uchl1 translation through an embedded SINEB2 repeat. Nature 491(7424):454–457. https://doi.org/10.1038/nature11508

5. Schein A, Zucchelli S, Kauppinen S, Gustincich S, Carninci P (2016) Identification of antisense long noncoding RNAs that function as SINEUPs in human cells. Sci Rep 6: 33605. https://doi.org/10.1038/srep33605

6. Zucchelli S, Cotella D, Takahashi H, Carrieri C, Cimatti L, Fasolo F, Jones MH, Sblattero D, Sanges R, Santoro C, Persichetti F, Carninci P, Gustincich S (2015) SINEUPs: a new class of natural and synthetic antisense long non-coding RNAs that activate translation. RNA Biol 12(8):771–779. https://doi.org/10.1080/15476286.2015.1060395

7. Zucchelli S, Fasolo F, Russo R, Cimatti L, Patrucco L, Takahashi H, Jones MH, Santoro C, Sblattero D, Cotella D, Persichetti F, Carninci P, Gustincich S (2015) SINEUPs are modular antisense long non-coding RNAs that increase synthesis of target proteins in cells. Front Cell Neurosci 9: 174. https://doi.org/10.3389/fncel.2015.00174

8. Patrucco L, Chiesa A, Soluri MF, Fasolo F, Takahashi H, Carninci P, Zucchelli S, Santoro C, Gustincich S, Sblattero D, Cotella D (2015) Engineering mammalian cell factories with SINEUP noncoding RNAs to improve translation of secreted proteins. Gene 569(2):287–293. https://doi.org/10.1016/j.gene.2015.05.070

9. Sasso E, Latino D, Froechlich G, Succoio M, Passariello M, De Lorenzo C, Nicosia A, Zambrano N (2018) A long non-coding SINEUP RNA boosts semi-stable production of fully human monoclonal antibodies in HEK293E cells. MAbs 10(5):730–737. https://doi.org/10.1080/19420862.2018.1463945

10. Zucchelli S, Patrucco L, Persichetti F, Gustincich S, Cotella D (2016) Engineering translation in mammalian cell factories to increase protein yield: the unexpected use of long non-coding SINEUP RNAs. Comput Struct Biotechnol J 14:404–410. https://doi.org/10.1016/j.csbj.2016.10.004

11. Espinoza S, Scarpato M, Damiani D, Manago F, Mereu M, Contestabile A, Peruzzo O, Carninci P, Santoro C, Papaleo F, Mingozzi F, Ronzitti G, Zucchelli S, Gustincich S (2020) SINEUP non-coding RNA targeting GDNF rescues motor deficits and neurodegeneration in a mouse model of Parkinson's disease. Mol Ther 28(2):642–652. https://doi.org/10.1016/j.ymthe.2019.08.005

12. Bon C, Luffarelli R, Russo R, Fortuni S, Pierattini B, Santulli C, Fimiani C, Persichetti F, Cotella D, Mallamaci A, Santoro C, Carninci P, Espinoza S, Testi R, Zucchelli S, Condo I, Gustincich S (2019) SINEUP non-coding RNAs rescue defective frataxin expression and activity in a cellular model of Friedreich's Ataxia. Nucl Acids Res 47(20):10728–10743. https://doi.org/10.1093/nar/gkz798

13. Indrieri A, Grimaldi C, Zucchelli S, Tammaro R, Gustincich S, Franco B (2016) Synthetic long non-coding RNAs [SINEUPs] rescue defective gene expression in vivo. Sci Rep 6:27315. https://doi.org/10.1038/srep27315

14. Sheridan SD, Theriault KM, Reis SA, Zhou F, Madison JM, Daheron L, Loring JF, Haggarty SJ (2011) Epigenetic characterization of the FMR1 gene and aberrant neurodevelopment in human induced pluripotent stem cell models of fragile X syndrome. PLoS One 6(10): e26203. https://doi.org/10.1371/journal.pone.0026203

15. Severin J, Lizio M, Harshbarger J, Kawaji H, Daub CO, Hayashizaki Y, Consortium F, Bertin N, Forrest AR (2014) Interactive visualization and analysis of large-scale sequencing datasets using ZENBU. Nat Biotechnol 32(3):217–219. https://doi.org/10.1038/nbt.2840

16. Yao Y, Jin S, Long H, Yu Y, Zhang Z, Cheng G, Xu C, Ding Y, Guan Q, Li N, Fu S, Chen XJ, Yan YB, Zhang H, Tong P, Tan Y, Yu Y, Fu S, Li J, He GJ, Wu Q (2015) RNAe: an effective method for targeted protein translation enhancement by artificial non-coding

RNA with SINEB2 repeat. Nucl Acids Res 43(9):e58. https://doi.org/10.1093/nar/gkv125

17. Takahashi H, Kozhuharova A, Sharma H, Hirose M, Ohyama T, Fasolo F, Yamazaki T, Cotella D, Santoro C, Zucchelli S, Gustincich S, Carninci P (2018) Identification of functional features of synthetic SINEUPs, antisense lncRNAs that specifically enhance protein translation. PLoS One 13(2): e0183229. https://doi.org/10.1371/journal.pone.0183229

18. Toki N, Takahashi H, Zucchelli S, Gustincich S, Carninci P (2020) Synthetic in vitro transcribed lncRNAs (SINEUPs) with chemical modifications enhance target mRNA translation. FEBS Lett 594(24):4357–4369. https://doi.org/10.1002/1873-3468.13928

19. Gruber AR, Lorenz R, Bernhart SH, Neubock R, Hofacker IL (2008) The Vienna RNA websuite. Nucl Acids Res 36(Web Server Issue):W70–W74. https://doi.org/10.1093/nar/gkn188

20. Lorenz R, Bernhart SH, Honer Zu Siederdissen C, Tafer H, Flamm C, Stadler PF, Hofacker IL (2011) ViennaRNA Package 2.0. Algor Mol Biol 6:26. https://doi.org/10.1186/1748-7188-6-26

21. Segundo-Val IS, Sanz-Lozano CS (2016) Introduction to the gene expression analysis. Methods Mol Biol 1434:29–43. https://doi.org/10.1007/978-1-4939-3652-6_3

22. Takahashi H, Sharma H, Carninci P (2019) Cell based assays of SINEUP non-coding RNAs that can specifically enhance mRNA translation. J Vis Exp (144). https://doi.org/10.3791/58627

23. Kozak M (2002) Pushing the limits of the scanning mechanism for initiation of translation. Gene 299(1–2):1–34. https://doi.org/10.1016/s0378-1119(02)01056-9

24. Flynn RA, Zhang QC, Spitale RC, Lee B, Mumbach MR, Chang HY (2016) Transcriptome-wide interrogation of RNA secondary structure in living cells with icSHAPE. Nat Protoc 11(2):273–290. https://doi.org/10.1038/nprot.2016.011

25. Ohyama T, Takahashi H, Sharma H, Yamazaki T, Gustincich S, Ishii Y, Carninci P (2020) An NMR-based approach reveals the core structure of the functional domain of SINEUP lncRNAs. Nucl Acids Res 48(16):9346–9360. https://doi.org/10.1093/nar/gkaa598

16

Development and use of Cellular Systems to Assess and Correct Splicing Defects

Nuria Suárez-Herrera, Tomasz Z. Tomkiewicz, Alejandro Garanto and Rob W. J. Collin

Abstract

A significant proportion of mutations underlying genetic disorders affect pre-mRNA splicing, generally causing partial or total skipping of exons, and/or inclusion of pseudoexons. These changes often lead to the formation of aberrant transcripts that can induce nonsense-mediated decay, and a subsequent lack of functional protein. For some genetic disorders, including inherited retinal diseases (IRDs), reproducing splicing dynamics in vitro is a challenge due to the specific environment provided by, e.g. the retinal tissue, cells of which cannot be easily obtained and/or cultured. Here, we describe how to engineer splicing vectors, validate the reliability and reproducibility of alternative cellular systems, assess pre-mRNA splicing defects involved in IRD, and finally correct those by using antisense oligonucleotide-based strategies.

Key words *ABCA4*, Antisense oligonucleotide, Exon skipping, Genetic therapy, Inherited retinal diseases, Maxigene, Midigene, Pre-mRNA, Pseudoexon, Splicing modulation, Splicing vectors

1 Introduction

Technologies such as next generation sequencing (NGS) expanded the discovery of genetic variants from coding regions to the entire genome. As a consequence of high-throughput data analysis, it is crucial to be able to correctly identify and distinguish disease-causing variants from single nucleotide polymorphisms (SNPs). This is especially relevant for intronic variants, as many of them have an unknown functional significance. In the field of inherited retinal diseases (IRDs), a common autosomal recessive condition known as Stargardt disease (STDG1) [1] lacks the bi-allelic molecular diagnosis in 30% of cases, i.e. the second variant cannot be identified in the coding regions of adenosine triphosphate (ATP) binding cassette type A4 (*ABCA4*) gene [2, 3].

Nuria Suárez-Herrera and Tomasz Z. Tomkiewicz contributed equally to this work.

According to the Human Gene Mutation Database, mis-splicing mutations have been estimated to represent 8.6% of the total mutations underlying inherited diseases (23,868/ 275,716) [4]. In vitro functional assays can provide insight into the underlying mechanisms behind aberrant splicing and identify the mutations that interfere with this process. Currently, the ideal model to assess and correct mis-splicing mutations in STGD1 are iPSC-derived retina-like cells (such as photoreceptor precursor cells (PPCs) or retinal organoids) from a patient harboring the genetic variants of interest, as these represent the right cell type(s) with the proper genetic context. In parallel, great efforts have been made to develop a more cost-effective and less time-consuming strategy to reach the same goal, by trying to mimic the pathological situation in a reliable and controlled manner. Engineering and use of multi-exon splice vectors have been proven to be extremely effective [2, 5–7] in gaining insight into IRDs and, more specifically, STDG1. In general, splice vectors or midigenes refer to a specific type of vector containing a large genomic region that allows to study the splicing processes between the included exons. An even longer genomic content allows for inclusion of long-range cis-acting elements to more accurately reflect the dynamics of splicing [8]. These "artificial" genomic vectors were shown to be very valuable when it comes to *ABCA4*, as the entire 128-kb gene has been successfully spanned in a set of midigenes representing an alternative to the impossibility of cloning such a large genomic region in a single vector [6].

The suitability of midigenes for some cell lines reduced the complexity of the study of pathologic deep-intronic variants (among other types of mutations) and their effect on splicing. In addition, validated midigenes harboring splicing variants represent a system for reliable and relatively quick identification of potential therapeutic molecules such as antisense oligonucleotides (AONs). AON-based therapies represent a very effective approach to target mis-splicing mutations. AONs are chemically modified RNA molecules that have the ability of modulating splicing by binding to pre-mRNA and interfering with the spliceosome [9]. They are used in the field of IRDs [10], as well as in other genetic diseases as the purpose of AONs is not limited to splice-switch function only [11]. Generating a reliable artificial splicing system is of major importance when it comes to the development of a potential therapeutic molecule. For the early investigation of possible causative variant affecting pre-mRNA splicing in retinal genes and subsequent assessment of AON potency, midigene technology is a suitable approach that has been shown to produce reliable results [6, 12].

Following transfection of the vector into HEK293T cells, AONs are co-transfected, and the splicing correction can be assessed at RNA level. Subsequently, the AONs that are identified as most potent can be tested in more advanced and precise cell models such as iPSC-derived retina-like cells.

Despite the proven efficiency of the HEK293T cells, they are not derived from ocular tissue, which allows to speculate whether the splicing dynamics enforced by HEK293T represents that of the retina. As an alternative, with the aim to better represent the retina-like splicing dynamics in in vitro splice assays, we also describe the use of retinoblastoma WERI-Rb-1 cells. These cells are also suitable for midigene transfection and in some cases demonstrate splicing dynamics that are more similar to that of the retina when compared to HEK293T cells.

In this chapter, we describe how to design multi-exon splice vectors followed by appropriate validation and correction of mis-splicing mutations by performing in vitro studies in cellular systems.

2   Materials

2.1   Design of Midigene and Maxigene Splice Vector

1. Donor and destination vector (*see* **Note 1**).

2. BP-Clonase cloning kit: BP-Clonase, buffer, proteinase K, etc. (i.e. Gateway® enzymes).

3. LR-Clonase cloning kit: LR-Clonase, buffer, proteinase K, etc. (i.e. Gateway® enzymes).

4. Genomic DNA, in this example, the bacterial artificial chromosome (BAC) clone, CH17-325O16 (insert g.94,434, 639–94,670,492), containing the entire *ABCA4* gene.

5. Generated midigenes (*see* Subheading 3.1.1).

6. Primers flanking the region of interest with attB sites. In this chapter we use *ABCA4* as an example. Forward primer sequence (attB sites underlined): 5′-<u>GGGGACAAGTTTGT-ACAAAAAAGCAGGCTTC</u> aacactgctggcaattggag-3′ and reverse primer sequence: 5′-<u>GGGGACCACTTTGTACAA GAAAGCTGGGTG</u> agctactgtgtggagggtg-3′. Primers are located in intron 6 and intron 11 of *ABCA4*, and serve as an example [6].

7. In silico cloning software (VectorNTI, SnapGene or Benchling).

8. High-fidelity *Taq* polymerase PCR kit: High-fidelity *Taq* polymerase, dNTPs, buffer, $MgCl_2$ and Q-solution or DMSO if applicable.

9. Unique restriction enzymes for sites present in your constructs, in our example are shown as A and B (*see* Subheading 3.2.1).

10. Commercially available DNA purification and cleanup kit.

11. Shrimp Alkaline Phosphatase.

12. T4 ligation kit.

13. Competent cells (preferably commercial ones).

14. LB medium: Autoclave 10 g NaCl, 10 g tryptone, and 5 g yeast extract in 1 L of deionized water.

15. LB plates: Autoclave 10 g NaCl, 10 g tryptone, 5 g yeast extract, and 30 g agar in 1 L of deionized water.

16. Selection antibiotics (usually ampicillin and kanamycin) at 50 mg/mL (this is the stock 1000× concentrated).

17. Commercially available Mini/Midiprep kit for plasmid DNA purification.

18. Electrophoresis equipment and agarose gels.

2.2 Site-Directed Mutagenesis

1. Generated midigene and maxigene vectors (*see* Subheadings 3.1.1 and 3.2.2).

2. Primers to introduce the desired mutation, in this chapter we use the c.859-506G > C mutation in the *ABCA4* midigene as an example (the variant is in bold and underlined):

    Forward primer 5′- CTGTGATTTGTTGTTGTTGTTG TTGTTGTTTT **G** AGACGGAGTAT -TGCTCAG-3′ and reverse primer 5′- GACACTAAACAACAACAACAACAACAA CAA-AA**C**TCTGCCTCATAACGAGTC-3′ [3].

3. Primers to amplify the region within selected restriction sites of the midi-/maxigene.

4. High-fidelity *Taq* polymerase PCR kit: High-fidelity *Taq* polymerase, dNTPs, buffer, $MgCl_2$ and Q-solution or DMSO if applicable.

5. Standard *Taq* polymerase kit.

6. pGEM®-T Easy Vector System kit.

7. IPTG and X-Gal.

8. EcoRI restriction enzyme.

9. DpnI restriction enzyme.

10. Competent cells (preferably commercial ones).

11. LB medium: Autoclave 10 g NaCl, 10 g tryptone, and 5 g yeast extract in 1 L of deionized water.

12. LB plates: Autoclave 10 g NaCl, 10 g tryptone, 5 g yeast extract, and 30 g agar in 1 L of deionized water.

13. Selection antibiotics (usually Ampicillin and Kanamycin) at 50 mg/mL.

14. Commercially available Mini/Midiprep kit for plasmid DNA purification.

15. Electrophoresis equipment and agarose gels.

16. Corresponding restriction enzymes C and D (*see* Subheading 3.2.3).

17. Shrimp Alkaline Phosphatase.

18. Commercially available DNA purification and cleanup kit.

19. T4 ligation kit.

2.3 Culture Conditions and Cell Lines

1. HEK293T cells (ATCC® CRL-3216™). Culture medium: DMEM 10% FCS medium (DMEM medium supplemented with 10% Fetal Calf Serum (FCS), 100 U/mL of penicillin, 100 μg/mL streptomycin, and 1% (v/v) 100 mM sodium pyruvate).

2. WERI-Rb-1 cells (ATCC® HTB-169™). Culture medium: DMEM 15% FCS medium (DMEM medium supplemented with 15% FCS, 100 U/mL of penicillin, 100 μg/mL streptomycin and 10 mL of 1 M HEPES).

3. T75 flasks to culture cell lines.

4. 0.25% trypsin solution for cell dissociation.

5. 1× PBS.

2.4 Midigene and AON Transfection

1. HEK293T or WERI-Rb-1 cells and corresponding culture medium.

2. Midigene vector, as an example we used *ABCA4* BA7 midigene [6].

3. AON stock: resuspend the lyophilized AONs at final concentration of 0.1–1 mM in 1× PBS previously autoclaved twice.

4. 6-well plates and 24-well plates.

5. OptiMEM and transfection reagents (i.e., FuGene® or Lipofectamine®).

6. 0.25% trypsin solution for cell dissociation.

7. 1× PBS.

2.5 RT-PCR

1. Commercially available RNA isolation kit.

2. Commercially available cDNA synthesis kit.

3. Primers located in the flanking exons of your region of interest. In the example described in this chapter:

(a) Region of interest.

- *ABCA4* exon 7 forward: 5'- TCTGAGATCTTGGG GAGGAA-3'.

- *ABCA4* exon 8 reverse: 5'- TGGAGTCAATCCCCA GAAAG-3'.

(b) Actin loading control (*see* **Note 2**).

- *ACTB* exon 3 forward: 5′-ACTGGGACGACATGGA GAAG-3′.

- *ACTB* exon 4 reverse: 5′-TCTCAGCTGTGGTGGT GAAG-3′.

(c) *RHO* transfection control.

- *RHO* exon 5 forward: 5′-ATCTGCTGCGGCAA GAAC-3′.

- *RHO* exon 5 reverse: 5′-AGGTGTAGGGGATGGGA GAC-3′.

4. PCR kit: Polymerase, dNTPs, buffer, $MgCl_2$, water, and Q-solution or DMSO if applicable.

5. Electrophoresis equipment and agarose gels.

3 Methods

Retina-specific genes are not readily expressed outside the ocular tissue. The inability to express or poorly express the genes of interest in non-ocular tissues makes it difficult to study variants affecting pre-mRNA splicing. Generation of PPCs from patient-derived reprogramed fibroblasts is an alternative to this, but is time-consuming and expensive.

3.1 Design of Midigene Splice Vectors

3.1.1 Gateway Cloning

In here, we describe the generation and use of pCI-NEO-*RHO* Gateway-adapted in-house vector (*see* Fig. 1).

1. Identify the gene and sequence of interest in genomic databases such as Ensembl Genome Browser or UCSC [13, 14] and then identify the region of interest. In this case, the region of interest is a deep-intronic variant causing guanine to cytosine substitution at position c.859-506 of the *ABCA4* gene. The base substitution strenghtes a cryptic deep-intronic splice acceptor side and causes generation of a 56-nts long pseudoexon between exon 7 and 8.

2. Design primers suitable for Gateway® BP cloning (*see* **Note 3**).

3. Use the BAC clone, CH17-325O16 (insert g.94,434, 639–94,670,492), containing the entire *ABCA4* gene. Isolate the BAC DNA using commercially available midiprep kit and use it as a PCR template to generatet the midigenes. Prepare the PCR reaction with 0.5 μM of each forward and reverse primer, 0.2 mM dNTPs, Phusion High-Fidelity DNA polymerase, 1× Phusion GC buffer, 3% DMSO, and 2.5 ng of BAC DNA in a total of 50 μL. Run the PCR program where the initial denaturation is at 98 °C for 30 s; 15–20 cycles of denaturation at 98 °C for 10 s each, annealing at 58 °C and

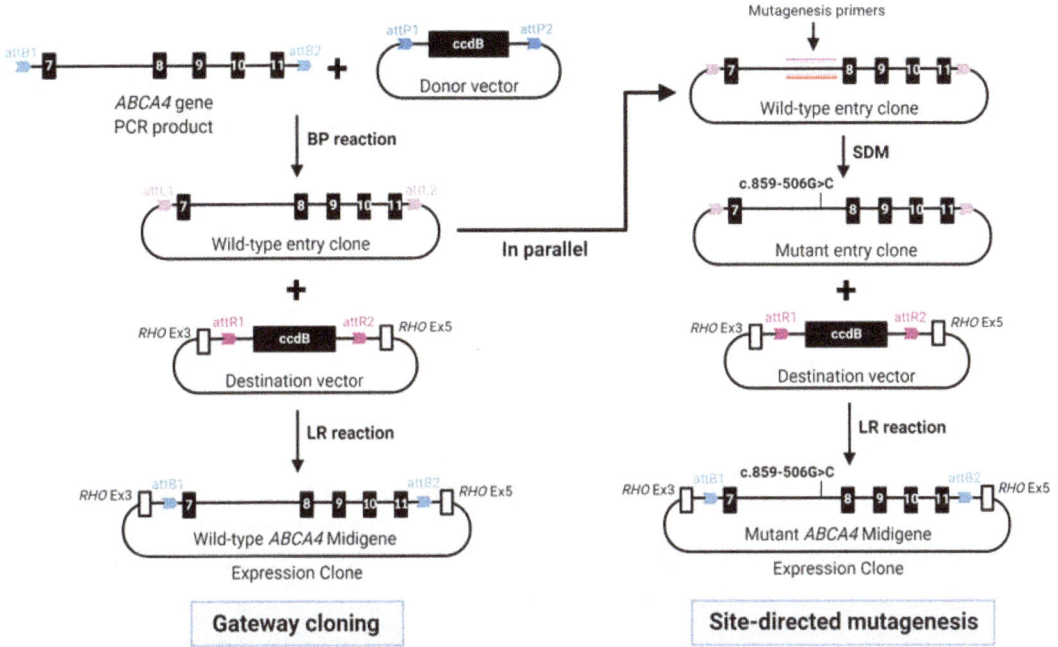

Fig. 1 Schematic representation of wild-type and mutant *ABCA4* midigene engineering. The simplified protocol for Gateway® system cloning and site-directed mutagenesis are shown on the left and right section, respectively. *SDM*: site-directed mutagenesis

extension at 72 °C for at least 1 min per kb of insert, with a final extension at 72 °C for 15 min [6].

4. Resolve the PCR product by gel electrophoresis by loading 10% of the reaction. The presence of a single band at a corresponding size indicates a succesful PCR amplification step. The band needs to be purified using any available commercial DNA purification and cleanup kit.

5. Set-up the BP reaction as instructed by the manufacturer of the BP-clonase cloning kit. The procedure used in-house includes 1 μL of donor vector (150 ng), 2 μL of buffer, 150 ng of purified and sequenced PCR product (max. 5 μL), milli-Q water up to 8 and 2 μL of BP-Clonase enzyme. The BP reaction has to be incubated at 25 °C for a minimum of 2 h.

6. Terminate the reaction by adding 2 μL of Proteinase K and incubating for 10 min at 37 °C.

7. Transform up to 5 μL of the reaction using competent cells (*see* **Note 4**) and incubate for 30 min on ice.

8. Perform the heat shock between 45 and 60 s at 42 °C and immediately place the cells back on ice for a minimum of 2 min.

9. Add 250 μL of SOC medium into the tube and incubate for 1 h at 37 °C.

10. Plate the content of the tube on a plate containing the corresponding antibiotic (the in-house vector has a kanamycin-resistance cassette) and incubate O/N at 37 °C.

11. Pick between 5 and 10 colonies and grow them in a 3 mL of LB medium supplemented with the corresponding antibiotic (1:1000 ratio of antibiotic to medium) O/N.

12. Perform the plasmid isolation using commercially available miniprep kit for plasmid DNA purification.

13. Verify the presence of insert by conducting restriction analysis. Use donor vector as a control (*see* **Note 5**).

14. Sequence the entire wild-type clone to make sure that the *Taq* did not introduced new mutations during the amplification step (*see* **Note 6**).

15. Perform side-directed mutagenesis (*see* Subheading 3.1.2).

3.1.2 Side-Directed Mutagenesis

In this section, we describe the side-directed mutagenesis protocol for midigene constructs:

1. Design the mutagenesis primers to have approximatelly 20-nts flanking both regions of the c.859-506 position (*see* Fig. 1).

2. Prepare mutagenesis mastermix with 1 U of high-fidelity *Taq* polymerase, 0.5 µM of forward and reverse primer, 0.2 mM dNTPs, 1× high-fidelity reaction buffer, 2× Q-solution, and 10–35 ng of the wild-type midigene vector in a total of 50 µL. Run the PCR program where the initial denaturation is at 94 °C for 5 min; 15 cycles of denaturation at 94 °C for 30 s each, annealing between 50 and 58 °C for 30 s and extension time at least 1 min per each kb of the complete plasmid with a final extension at 72 °C for 20 min.

3. Add 1 µL DpnI directly to 20 µL of the PCR reaction to digest the original template (the wild-type sequence). Incubate the reaction at 37 °C for 3 h. The reaction is terminated at 80 °C for 20 min.

4. Perform the transformation using up to 5 µL of the reaction (*see* **steps 7–13** in Subheading 3.1.1).

5. Sequence the entire mutant clone to confirm the presense of the desired mutation and also to identify all of the undesired substitutions present in the mutant clone (*see* **Note 6**).

3.2 Design of Maxigene Splice Vectors

There are some variants that may need larger genomic environment in order to correctly assess them. This could be the case for mutations whose effect is only detected when other splice regulatory motifs are present, generally located in introns. As a consequence, larger genomic context should be included and based on our experience, it is difficult to obtain a splicing vector of this size because of several reasons. The first one is the probability of inducing single nucleotide changes during the amplification of the genomic region of interest. Long-range PCR and high-fidelity DNA *Taq* polymerases may prevent this from happening, although these chances increase when amplifying >10 kb fragments. Recombination efficiency between the fragment and the donor vector is also affected, as well as the efficiency of site-directed mutagenesis on the entry clone. Both cases are highly associated with the size of the vector. To overcome these limitations, we propose to use the already engineered midigenes completely covering the whole gene in order to generate a maxigene, which combines the genomic context of more than one midigene. In this section, we show an example of the mentioned cloning strategy.

3.2.1 In Silico Design of Maxigene Strategy

1. Select the two midigenes that include the introns and exons of interest.

2. Find a common region within both vectors (generally, it is the overlapping sequence between the first midigene and the following).

3. Look for restriction sites in the common region (A) as well as in the backbone of the vector (B) (*see* Fig. 2). Restriction sites A and B should be unique in both midigenes (*see* **Note 7**).

4. Simulate the engineering of the final construct by selecting the fragments flanked by restriction sites A and B in both midigenes. From the first vector (based on sequence), select the fragment B → A (backbone to common region), whereas in the second vector the fragment goes A → B (common region to backbone).

3.2.2 Cloning of Maxigene Vectors

1. Check if the enzymes are compatible in terms of incubation time/temperature, reaction buffer and stability in order to avoid partial digestions (*see* **Note 8**).

2. Set the digestion reaction for midigene 1 and 2 with restriction enzymes A and B (*see* Fig. 2). Digest at least 0,5–1 µg of DNA and check the digestion product by gel electrophoresis (*see* **Note 9**).

3. Cut the band corresponding to the fragment of interest and purify the DNA by using the kit of your convenience (*see* **Note 10**).

4. Measure DNA concentration and perform the ligation reaction as described by the manufacturer of the T4 ligase kit (make sure to include a negative control the vector only).

5. Transform 5 µL of the ligation product using DH10β competent cells (*see* **Note 4**) and incubate for 20–30 min on ice.

6. Perform the heat shock for 30 s at 42 °C and cool down the cells for 2 min on ice.

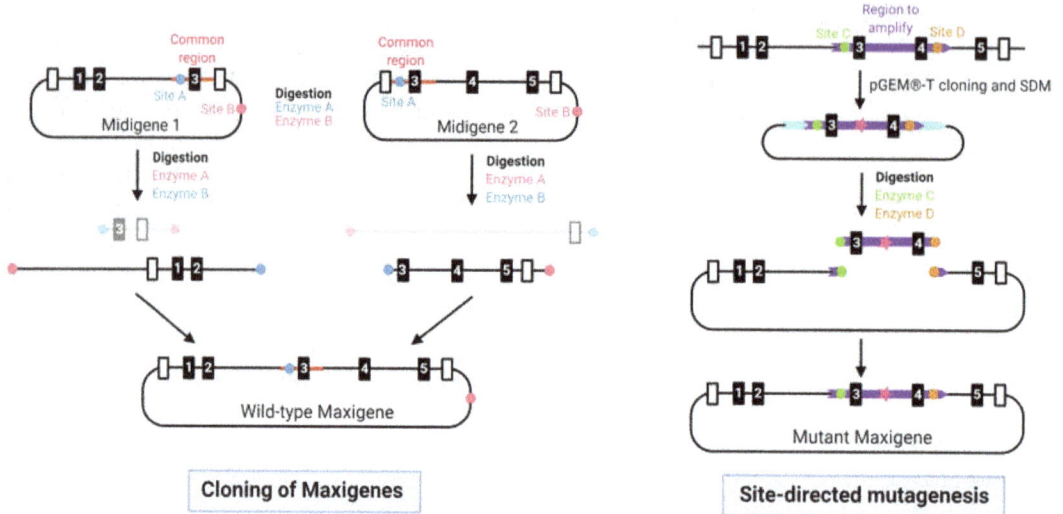

Fig. 2 Schematic representation of wild-type and mutant maxigene engineering. Cloning steps and site-directed mutagenesis are shown on the left and right section, respectively. This example of maxigene strategy is based on the alternative procedures explained in Subheading 3.2. *SDM*: site-directed mutagenesis

7. Add 250 µL of 10β/Stable Outgrowth Medium into the tubes and incubate for 1 h at 37 °C in the shaking incubator (250 rpm).

8. Plate everything on LB-agar plates containing the corresponding antibiotic (as we are referring to expression clones, these have ampicillin resistance).

9. Incubate O/N at 37 °C.

10. Take the plates from the incubator in the morning and in the same afternoon, pick colonies and let them grow at 37 °C in 3 mL of LB medium supplemented with the corresponding antibiotic, i.e., ampicillin.

11. Perform plasmid isolation and verify if the ligation worked by restriction analysis or colony PCR. As a negative control, use midigene 1 in parallel.

12. Sequence the positive clones in order to verify that they do not contain any additional mutations.

3.2.3 Site-Directed Mutagenesis

To obtain the mutant maxigene, you can either use midigene 1 or 2 containing the mutation of interest and follow the same protocol indicated in the previous section or perform the site-directed mutagenesis on the wild-type maxigene. In order to perform the second option, we propose the following strategy as an example for a maxigene (*see* Fig. 2):

1. Look for unique restriction sites flanking the region of interest (containing the position where single-nucleotide change has to be performed, in this case would be C and D).

2. Amplify this region by using high-fidelity PCR kit.

3. Incubate with normal *Taq* Polymerase for 20 min at 72 °C to add the A overhangs to the final PCR product.

4. Clone the fragment into a pGEM®-T vector following the instructions of the kit's manufacturer. Incubate O/N at 4 °C.

5. The next day, transform 5 μL of the previous reaction using DH5α competent cells.

6. Perform the heat shock for 60 s at 42 °C and cool down the cells for 2 min on ice.

7. Add 250 μL of stable outgrowth medium into the tubes and incubate for 1 h at 37 °C in the shaking incubator (250 rpm).

8. Just before plating, add IPTG and X-Gal in the tube and immediately, plate everything on LB-agar plates containing the corresponding antibiotic (pGEM®-T vectors have ampicillin resistance).

9. Incubate O/N at 37 °C.

10. Pick white colonies and let them grow at 37 °C in 3 mL of LB medium supplemented with antibiotic.

11. Perform plasmid isolation using available commercial miniprep kit for plasmid DNA purification, followed by restriction analysis using EcoRI to check if the region is cloned. Keep in mind that your insert may contain an EcoRI restriction site.

12. Sequence the positive clones to check if they contain any additional mutations.

13. Perform site-directed mutagenesis as previously indicated for midigene vectors.

14. Verify the mutation by Sanger sequencing.

15. Digest both the pGEM®-T vector and the maxigene with restriction enzymes C and D.

16. Set the ligation reaction as indicated in the maxigene cloning with either the purified digestion products directly (incubating one of the fragments with phosphatase) or the purified fragments from the agarose gel.

17. Pick colonies and perform restriction analysis or colony PCR to check the correct ligation between the fragment containing the mutation and the rest of the maxigene.

18. Analyze the new inserted region by Sanger sequencing.

3.3 In Vitro Evaluation of Splice Vectors in Cell Lines

The constructed midi/maxigenes need to be validated and assessed in vitro. In the functional studies of IRDs, the cell line of choice is usually HEK293T. These cells are easy to transfect and do not express retina-specific genes. Therefore, the expression of retina-specific gene delivered with the vector is not interfered by the endogenous expression of such gene. This advantage is counterbalanced by fact that the HEK293T cells have different properties compared to retina cells. As a consequence, the splicing dynamics are often but not always the same. As an alternative to better represent the retina-like splicing dynamics, we also describe the use of WERI-Rb-1 cells, which are retinoblastoma cells. These cells are still suitable for midigene transfection and sometimes mimic retina-specific splicing patterns better in comparison with HEK293T cells (*see* Fig. 3).

3.3.1 Transfection in HEK293T

1. Seed 0.5×10^6 cells/well in DMEM 10% FCS medium in 6-well plate if you can transfect the midigene 4 h post-seeding (*see* **Note 11**).

2. Once the cells are attached, transfect 1.2 μg of plasmid in a 6-well plate. In this example we used FuGENE (ratio is 3:1 FuGENE:plasmid). To make transfection mix, to 200 μL of OptiMEM medium, add 3.6 μL of FuGENE. To the transfection mix add 1.2 μg of plasmid and incubate at RT for 15–20 min.

3. Dispense the transfection mix on top of the corresponding well (*see* **Note 12**).

4. Incubate the plate at 37 °C for 24 (*see* Subheading 3.4.1) or 48 h (*see* **Note 13**).

5. After the incubation, gently rinse the cells with pre-warmed 1× PBS and then detach the cells using 500 μL trypsin. Collect the detached cells into a 1.5 mL Eppendorf using P1000 pipette.

6. Centrifuge the Eppendorf tubes for 5 min at $1000 \times g$ to pellet the cells, remove the supernatant and wash the cells again in 1× PBS for 5 min at $1000 \times g$.

7. Discard the supernatant and:

   (a) freeze the cell pellets at −80 °C (storage) after 48 h incubation and proceed with further analysis at another moment or,

   (b) proceed with the RNA isolation (validation of the midigene expression and splicing) after 48 h incubation.

3.3.2 Transfection in WERI-Rb-1

1. Dispense 1 to 2×10^6 cells to 1.5 mL Eppendorf tube in a total volume of 500 μL of DMEM 15% FCS.

2. Transfect 1.2 μg of plasmid. In this example, we used FuGENE (ratio is 3:1 FuGENE:plasmid). To make transfection mix

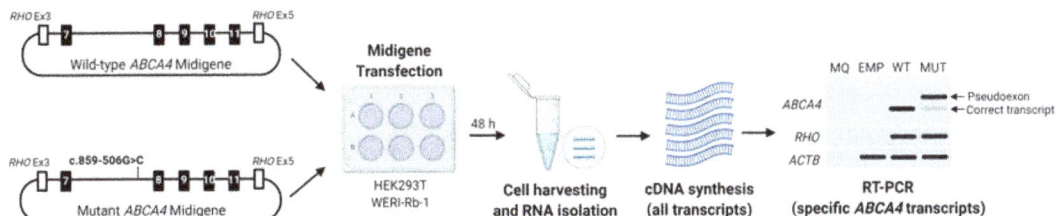

Fig. 3 Experimental strategy to assess splicing defects in HEK293T or WERI-Rb-1 cell lines. The cells are first transfected with a wild-type (WT) or mutant (MUT) *ABCA4* midigene and, following a 48-h incubation, splicing is validated by RT-PCR as outlined in Subheading 3.4.3 The gel picture shows an approximate read-out of the expected *ABCA4* transcripts. *MQ* negative control of the PCR reaction, *EMP* empty transfection mix (endogenous expression of the selected genes within the cell line used)

200 μL of OptiMEM medium with 3.6 μL of FuGENE. And to this tube, add 1.2 μg of plasmid and incubate at RT for 15–20 min.

3. Dispense the transfection mix to 1.5 mL Eppendorf tube with WERI-Rb-1 cells and incubate for 2 h at 37 °C.

4. After the incubation, transfer the cells to the corresponding well on the 6-well plate and add medium to a total volume of 2 mL.

5. Incubate the plate at 37 °C for either 24 (*see* Subheading 3.4.1) or 48 h (*see* **Note 13**).

6. After incubation, collect the cells in 1.5 mL Eppendorf tube and spin it for 5 min at $1000 \times g$.

7. Remove the medium and wash the cells again in $1\times$ PBS for 5 min at $1000 \times g$.

8. Discard the PBS and:

   (a) freeze the cell pellets at -80 °C (storage) after 48 h incubation and proceed with further analysis at another moment or,

   (b) proceed immediately with the RNA isolation (validation of the midigene expression and splicing) after 48 h incubation.

3.3.3 Validation of Splicing Events by Reverse Transcriptase PCR (RT-PCR)

Before starting to use midigenes as an artificial system, it is necessary to check the effect of the variant at pre-mRNA level, and therefore, the RT-PCR is used as a validation method for splicing events.

1. Design forward and reverse primers to flank the region of interest. If the cell line used presents endogenous expression of your target gene, design one of the primers in the artificial exon of the midigene (*RHO*) (*see* **Note 14**).

2. Isolate the RNA from the pellets obtained in Subeading 3.3.1 or 3.3.2 and measure the concentration using NanoDrop.

3. Use 1 μg of RNA to synthesize cDNA by following the instruction of the kit's manufacturer.

4. Prepare RT-PCR master mix with 1 U *Taq* polymerase, 1× PCR buffer with $MgCl_2$, 0.2 mM dNTPs, and 50–60 ng of cDNA in a total volume of 25 μL. In parallel, prepare the corresponding PCR mixes for actin (loading control) and *RHO* (midigene transfection control) using the mastermix recipe outlined above with appropriate actin and *RHO* primers. Run the PCR program where the initial denaturation is at 94 °C for 5 min; 35 cycles of denaturation at 94 °C for 30 s each, annealing at 58 °C for 30 s and extension time at 72 °C for 3 min, with a final elongation step at 72 °C for 5 min.

5. Resolve PCR products by gel electrophoresis. Load 10 μL from *ABCA4* PCR products and 5 μL from actin and *RHO* PCR products as well.

6. Excise the bands coming from each midigene transfection and extract the DNA using a PCR cleanup kit of choice.

7. Measure DNA concentration and if it is not sufficient for Sanger sequencing (minimum of 200 ng per sample), clone the bands in pGEM®-T vector following the manufacturer's instructions.

8. Verify all bands corresponding to different splicing events comparing the mutant to the wild-type condition.

9. Correctly identify and remove midigene artifacts (*see* **Note 15**) or heteroduplexes (*see* **Note 16**).

3.4 Correcting Splicing Defects in Artificial Systems

In this part of the chapter, we describe the experimental design to correctly compare a set of AON sequences by using midigenes as a model system and how to determine the efficacy of aberrant splicing correction. The utilization of photoreceptor precursor cells and/or retinal organoids is a recommended step in the final stages of AON potency validation in in vitro studies.

3.4.1 Midigene Vector and AON Co-transfection

Seeding 0.5×10^6 cells/well in 6-well plate provides enough cells to seed 6 wells on the 24-well plate. It is recommended to perform early calculations into how many wells will be required for the splice correction assay as this will influence the number of wells on the 6-well plate required for midigene transfection. If 6 or less wells on the 24-well plate are needed, one 6-well plate seeded with 0.5×10^6 HEK293T cells and transfected with midigene is enough.

When setting up splice assay with AONs, do not forget about the minimum required controls. These include non-transfected control or empty transfection mix (EMP), cells transfected with

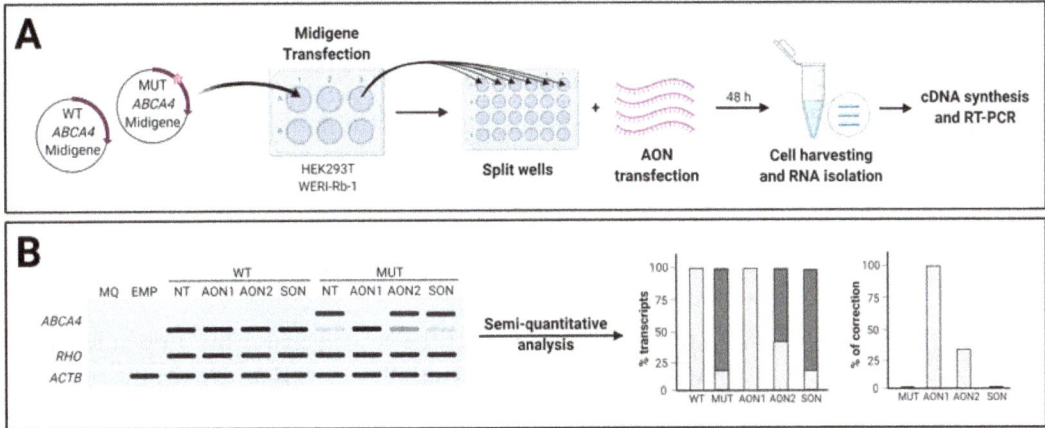

Fig. 4 Analysis of AON-mediated splicing correction in midigene-based splice assays. (**a**) Experimental strategy to correct splicing defects by co-transfection of wild-type (WT) or mutant (MUT) *ABCA4* midigenes with AONs in cell lines as described in Subheading 3.4.1. (**b**) Representation of splicing read-out on agarose gel and further semi-quantitative analysis. The gel picture represents an approximation of expected *ABCA4* transcripts, further semi-quantitative analysis and graphs are based on this estimation. *MQ*: negative control of the PCR reaction, *EMP*: empty transfection mix (endogenous expression of the selected genes within the cell line used), *NT*: non-treated (transfected with midigene but no AONs), *SON*: scrambled or sense oligonucleotide

wild-type (WT) midigene or mutant (MUT) midigene but not transfected with AONs (NT), and cells transfected with wild-type or mutant midigene and co-transfected with the corresponding scrambled or sense oligonucleotide (SON). The controls cells have to be seeded together with test cells and exposed to the exact same treatment conditions, which include medium change and transfer from 6-well plate to 24-well plate.

1. Refer to Subheadings 3.4.1 and 3.4.2 for cell seeding and midigene transfection procedures, then incubate for the next 24 h.

2. Use 500 μL of trypsin to detach the HEK293T. Neutralize the effect of trypsin with 500 μL of HEK293T specific medium and collect the cells in a 15 mL Falcon tube. Wash the well with extra 1 mL of medium to collect all the remaining cells. Add 1 mL of HEK293T medium to make 3 mL cell suspension and seed 500 μL of the suspension per well in the 24-well plate using P1000 pipette (*see* **Note 17**).

3. Place the 24-well plate in the 37 °C incubator and allow for at least 4 h incubation or until cells become attached to the wells.

4. In this example, the final concentration of the AON was at 0.5 μM (*see* **Note 18**). To co-transfect two wells on a 24-well plate (wild-type and mutant midigene) with the same AON at 0.5 μM, mix 50 μL of OptiMEM medium with 1 μL of FuGENE® . To this mixture add 5 μL of resuspended AON at

100 μM stock concentration. Incubate the transfection reaction for 15–20 min at RT. After incubation add enough medium to make 1 mL.

5. On a 24-well plate, remove all the medium from the transfected wells and gently dispense 500 μL of the transfection mix to the wells transfected with wild-type and mutant midigenes. (*see* **Note 19**).

6. Place the well plate back in the incubator and allow for 48 h incubation (*see* Fig. 4a).

7. After incubation, collect the cells in an 1.5 mL Eppendorf tube and centrifuge for 5 min at $1000 \times g$. Remove the medium and wash the cells again in $1 \times$ PBS for 5 min at $1000 \times g$.

8. Discard the PBS and

   (a) freeze the cell pellets at $-80\ °C$ (storage) and proceed with further analysis at another moment or,

   (b) proceed immediately with the RNA isolation and the cDNA synthesis following manufacturer's instructions.

9. After the cDNA synthesis perform RT-PCR as previously described (*see* Subheading 3.3.3).

10. Resolve PCR products by gel electrophoresis.

11. Similar strategy is used for WERI-Rb-1 transfection with AONs. First, refer to Subheading 3.3.3 for midigene transfection to WERI-Rb-1 cells, and follow the protocol outlined above.

3.4.2 Quantification of Splicing Redirection with Image J

Measuring efficiency of an AON-based strategy represents the last step in order to discard or select a potential therapeutic molecule for further clinical studies. There are several methods available to calculate this efficiency and, in this chapter, we focus on a semi-quantification strategy for RT-PCR readouts (*see* Fig. 4b).

1. Open the gel image (TIFF format) in Image J or Fiji (same software).

2. Use the rectangle tool to make selections of the bands you want to quantify (*see* **Note 20**).

3. Select the first lane and press Ctrl + 1.

4. Move the newly generated rectangle to the right and put it on top of the next lane, then press Ctrl + 2.

5. Repeat **step 4** as many times as samples you have in your gel image.

6. After selecting all bands, press Ctrl + 3 and a new window with all intensity peaks will appear.

7. Using the straight-line tool, draw a line at the bottom of every peak and close it in order to remove the background signal from the band intensity signal.

8. Use the wand tracing tool and click on the closed peak to measure its area. Automatically, a new window will appear showing area value while measuring all the peaks (*see* **Note 21**).

9. Copy all area values into an Excel sheet (replace dots by commas if necessary).

10. Calculate the total area of every condition. Use this value as a reference to calculate the % of correct transcript or aberrant transcript (*see* **Note 22**) as indicated below:

$$\% \text{ of Transcript} = \frac{\text{Area Transcript}}{\text{Total Area}} \times 100$$

11. Introduce the values in Graphpad Prism and calculate the average of the other replicates as well (*see* **Note 23**).

12. Create a graph of the % of correct/aberrant transcript for every condition.

13. Calculate the percentage of aberrant transcript decrease by setting the aberrant transcript of non-treated mutant midigene value as your reference, this will represent the % of correction.

14. Take the average of the decrease in all samples per groups and compare the data using a One-Way ANOVA for statistical analysis.

3.4.3 How to Know If an AON Is Effective?

Once you assess the aberrant transcript rescue of the different AONs that are being tested, it is possible to classify them in several groups depending on their splicing redirection efficiency (*see* **Note 24**). In previous studies [15], we classified AONs in the following five groups in order to analyze their properties in a straightforward way (*see* Fig. 4b): highly effective (>75% correction), effective (between 75% and 50% correction), moderately effective (between 50% and 25% correction), poorly effective (between 25% and 0% correction), and noneffective (when correction is not detected).

However, there are some limitations as this protocol is based on a semi-quantitative strategy, such as the interference of heteroduplexes and midigene artifacts. Even though we presented different ways of overcoming these limitations, there is the possibility of using other quantitative strategies [16]. As an example, Fragment Analyzer, TapeStation or digital droplet PCR can be implemented in order to get more accurate splicing readouts (*see* **Note 25**).

4   Notes

1. The donor vector used in this example was pDONR201 vector supplied from Invitrogen. The destination vector used was pCI-NEO-*RHO* vector, which was adapted to contain genomic region encompassing exons 3 through 5 of *RHO*.

2. In the example, we used in-house primers designed for mouse because of the high sequence conservation between mouse *Actb* and human *ACTB* gene.

3. Design the primers to capture as much of the *ABCA4* genomic content for the Gateway®-adapted vector cloning. The primers have to be able to insert *attB* sites for Gateway BP cloning.

4. Transformation efficiency for >15 kb constructs is significantly increased when using DH10β cells instead of DH5α.

5. Selection of enzyme(s) used to verify the insert depend on the backbone and the insert. We recommend to use an enzyme that cuts in at least two distinct places across the vector. If such enzyme cannot be found, we recommend to use one enzyme that cuts the insert and one enzyme that cuts the backbone of the vector.

6. If new undesired mutations are present within the insert, assess their potential effect on the splicing, in silico, by using tools designed to study pre-mRNA splicing (http://www.umd.be/HSF/HSF.shtml).

7. If that is not the case, uniqueness of restriction site A is not strictly needed for both vectors, but it is not recommended having more than two in one of the midigenes. On the contrary, restriction site B should always be unique in both cases. In our case, restriction site A was located twice in the midigene 1. For that reason, we amplified a part of midigene 1 including restriction site B until the beginning of the common region where the second restriction site A was (reverse primer included half of the restriction site A and phosphate group at 5′), then we used this product to proceed with the strategy.

8. If this is not the case, sequential digestion is also possible.

9. In our case, the amplified fragment from midigene 1 had to be digested with restriction enzyme B only as restriction site A was located twice in this plasmid, whereas midigene 2 needed sequential digestion with both enzymes A and B.

10. Based on our experience, even if agarose gels prepared with TAE buffer are recommended to obtain a better resolution of large fragments, the loss of DNA after purification with TAE-specific kits is very high. As an alternative, you can always do a cleanup directly from the digestion product, but

incubation with phosphatase is needed at least in one of the digested midigenes and screening of colonies afterwards might be more extensive. In our case, phosphatase incubation was performed in digestion product from midigene 2 to avoid re-ligation with itself.

11. Seed at lower density $(0.2$–0.25×10^6 cells/well) if you have to wait 24 h until transfection.

12. Dispense the transfection mix drop by drop across the well. After dispensing all the midi-/maxigenes to all the wells, gently swirl the plate to ensure homogenous distribution of the midigene.

13. Twenty-four hours incubation is recommended for AON co-transfection in transcript rescue experiments, 48 h incubation is recommended for assessing the expression of the wild-type midi-/maxigene or assessing the effect of mutation on the transcript.

14. Using *RHO* primers for the RT-PCR might reveal other splicing events induced by the artificial exons. However, you can reduce this artifact by performing nested PCR afterwards, substituting the *RHO* primer by one binding to the next available exon of your transcript.

15. Reduce the amount of transfected midigene if artifacts are masking other splicing events or change the combination of primers, narrowing the region of interest could reduce the amplification of these artifacts.

16. To remove heteroduplexes from the read out, we suggest set a 3–5 cycles PCR in the same conditions by adding 1 μL from the product obtained in **step 5**. However, this may affect the reliability of the method as one extra PCR step is added, for that reason it is important to validate the observed final read-out [17].

17. If you are working with more than one 6-well plate, pool all the cells transfected with the same midigene together. Calculate how much medium needs to be added. If you are working with less than 6 wells on the 24-well plate collect the cells in the total of 3 mL as described, seed the required number of wells and discard of the remaining cell suspension if not needed.

18. The AONs can be tested at varying molarities to assess the concentration at which AONs are most efficient. The concentration can vary between 0.1 and 1 μM.

19. As an alternative, it is also possible to remove from each well the exact volume of the transfection mix to be added, making sure the final concentration of AON does not change.

20. The rectangle size and position (in Y-axis) will be equal for all lanes and cannot be changed in further steps, so make sure to select a region that can cover all bands from the same lane.

21. Try to follow a consistent order through all the analysis, start measuring the band corresponding to the correct transcript and moving to the one of either pseudoexon inclusion (larger) or exon skipping (smaller) band for every different lane/ condition.

22. In the aberrant transcript band, total/partial pseudoexon inclusion or total/partial exon skipping can be counted as aberrant. If heteroduplexes were also included in the band quantification, half of the value of that band should be counted as correct transcript, whereas the other half should be included in the aberrant transcript.

23. It is recommended to repeat the same experiment at least three times (replicates) in order to perform further statistical analysis and make proper comparisons between all conditions

24. This is especially useful when screening a set of AONs with the aim of covering entire regions at pre-mRNA levels [15].

25. In a same study, you can include more than one splicing analysis and compare between them to obtain a more robust result.

Acknowledgments

This work was supported by European Union's Horizon 2020—Marie Sklodowska-Curie Actions grant no. 813490 (to R.W.J.C.) and Retina UK Foundation grant no. GR596 (to R.W.J.C.), Algemene Nederlandse Vereniging ter Voorkoming van Blindheid, Stichting Blinden-Penning, Landelijke Stichting voor Blinden en Slechtzienden, Stichting Oogfonds Nederland, Stichting Macula Degeneratie Fonds, and Stichting Retina Nederland Fonds (who contributed through UitZicht 2015-31 and 2018-21), together with the Rotterdamse Stichting Blindenbelangen, Stichting Blindenhulp, Stichting tot Verbetering van het Lot der Blinden, Stichting voor Ooglijders, and Stichting Dowilvo (to A.G. and R.W.J.C.). This work was also supported by the Foundation Fighting Blindness USA, grant no. PPA-0517-0717-RAD (to A.G. and R.W.J.C.). The funding organizations had no role in the design or conduct of this research. They provided unrestricted grants.

References

1. Tanna P et al (2017) Stargardt disease: clinical features, molecular genetics, animal models and therapeutic options. Br J Ophthalmol 101(1):25–30

2. Fadaie Z et al (2019) Identification of splice defects due to noncanonical splice site or deep-intronic variants in ABCA4. Hum Mutat 40(12):2365–2376

3. Sangermano R et al (2019) Deep-intronic ABCA4 variants explain missing heritability in Stargardt disease and allow correction of splice defects by antisense oligonucleotides. Genet Med 21(8):1751–1760

4. Stenson PD et al (2017) The human gene mutation database: towards a comprehensive repository of inherited mutation data for medical research, genetic diagnosis and next-generation sequencing studies. Hum Genet 136(6):665–677

5. Bauwens M et al (2019) ABCA4-associated disease as a model for missing heritability in autosomal recessive disorders: novel noncoding splice, cis-regulatory, structural, and recurrent hypomorphic variants. Genet Med 21(8):1761–1771

6. Sangermano R et al (2018) ABCA4 midigenes reveal the full splice spectrum of all reported noncanonical splice site variants in Stargardt disease. Genome Res 28(1):100–110

7. Garanto A et al (2016) In vitro and in vivo rescue of aberrant splicing in CEP290-associated LCA by antisense oligonucleotide delivery. Hum Mol Genet 25(12):2552–2563

8. Khan M et al (2019) Identification and analysis of genes associated with inherited retinal diseases. In: Weber BHF, Langmann T (eds) Retinal degeneration: methods and protocols. Springer, New York, pp 3–27

9. Hammond SM, Wood MJ (2011) Genetic therapies for RNA mis-splicing diseases. Trends Genet 27(5):196–205

10. Collin RW, Garanto A (2017) Applications of antisense oligonucleotides for the treatment of inherited retinal diseases. Curr Opin Ophthalmol 28(3):260–266

11. Sharma VK, Sharma RK, Singh SK (2014) Antisense oligonucleotides: modifications and clinical trials. MedChemComm 5(10):1454–1471

12. Tomkiewicz TZ et al (2021) Antisense oligonucleotide-based rescue of aberrant splicing defects caused by 15 pathogenic variants in ABCA4. Int J Mol Sci 22(9):4621

13. Cunningham F et al (2018) Ensembl 2019. Nucleic Acids Res 47(D1):D745–D751

14. Kent WJ et al (2002) The human genome browser at UCSC. Genome Res 12(6):996–1006

15. Garanto A et al (2019) Antisense oligonucleotide screening to optimize the rescue of the splicing defect caused by the recurrent deep-intronic ABCA4 variant c.4539+2001G>A in Stargardt disease. Genes 10(6):452

16. Hiller M et al (2018) A multicenter comparison of quantification methods for antisense oligonucleotide-induced DMD exon 51 skipping in Duchenne muscular dystrophy cell cultures. PLoS One 13(10):e0204485

17. Thompson JR, Marcelino LA, Polz MF (2002) Heteroduplexes in mixed-template amplifications: formation, consequence and elimination by 'reconditioning PCR'. Nucleic Acids Res 30(9):2083–2088

Permissions

All chapters in this book were first published by Springer; hereby published with permission under the Creative Commons Attribution License or equivalent. Every chapter published in this book has been scrutinized by our experts. Their significance has been extensively debated. The topics covered herein carry significant findings which will fuel the growth of the discipline. They may even be implemented as practical applications or may be referred to as a beginning point for another development.

The contributors of this book come from diverse backgrounds, making this book a truly international effort. This book will bring forth new frontiers with its revolutionizing research information and detailed analysis of the nascent developments around the world.

We would like to thank all the contributing authors for lending their expertise to make the book truly unique. They have played a crucial role in the development of this book. Without their invaluable contributions this book wouldn't have been possible. They have made vital efforts to compile up to date information on the varied aspects of this subject to make this book a valuable addition to the collection of many professionals and students.

This book was conceptualized with the vision of imparting up-to-date information and advanced data in this field. To ensure the same, a matchless editorial board was set up. Every individual on the board went through rigorous rounds of assessment to prove their worth. After which they invested a large part of their time researching and compiling the most relevant data for our readers.

The editorial board has been involved in producing this book since its inception. They have spent rigorous hours researching and exploring the diverse topics which have resulted in the successful publishing of this book. They have passed on their knowledge of decades through this book. To expedite this challenging task, the publisher supported the team at every step. A small team of assistant editors was also appointed to further simplify the editing procedure and attain best results for the readers.

Apart from the editorial board, the designing team has also invested a significant amount of their time in understanding the subject and creating the most relevant covers. They scrutinized every image to scout for the most suitable representation of the subject and create an appropriate cover for the book.

The publishing team has been an ardent support to the editorial, designing and production team. Their endless efforts to recruit the best for this project, has resulted in the accomplishment of this book. They are a veteran in the field of academics and their pool of knowledge is as vast as their experience in printing. Their expertise and guidance has proved useful at every step. Their uncompromising quality standards have made this book an exceptional effort. Their encouragement from time to time has been an inspiration for everyone.

The publisher and the editorial board hope that this book will prove to be a valuable piece of knowledge for researchers, students, practitioners and scholars across the globe.

List of Contributors

Michael J. Gait and Sudhir Agrawal

Paula Milán-Rois, Ciro Rodriguez-Diaz, Milagros Castellanos and Álvaro Somoza

Nerea Moreno, Irene González-Martínez, Rubén Artero and Estefanía Cerro-Herreros

Arístides López-Márquez, Ainhoa Martínez-Pizarro, Belén Pérez, Eva Richard and Lourdes R. Desviat

Elena Daoutsali and Ronald A. M. Buijsen

Virginia Arechavala-Gomeza and Alejandro Garanto

Jeroen Bremer and Peter C. van den Akker

Liliana Matos, Juliana I. Santos, Mª. Francisca Coutinho and Sandra Alves

Kwan-Leong Hau, Amelia Lane, Rosellina Guarascio and Michael E. Cheetham

Remko Goossens and Annemieke Aartsma-Rus

Valentina Palacio-Castañeda, Roland Brock and Wouter P. R. Verdurmen

Haiyan Zhou

Pablo Herrero-Hernandez, Atze J. Bergsma and W. W. M. Pim Pijnappel

Omer Aydin, Dilek Kanarya, Ummugulsum Yilmaz and Cansu Ümran Tunç

Michele Arnoldi, Giulia Zarantonello, Stefano Espinoza, Stefano Gustincich, Francesca Di Leva and Marta Biagioli

Nuria Suárez-Herrera, Tomasz Z. Tomkiewicz, Alejandro Garanto and Rob W. J. Collins

Index